SCHAUM'S *Easy* O

TRIGONOMETRY

Other Books in Schaum's Easy Outline Series Include:

SCHAUM'S *Easy* OUTLINES

TRIGONOMETRY

BASED ON SCHAUM'S *Outline of Trigonometry*
BY FRANK AYRES, JR.
AND ROBERT E. MOYER

ABRIDGEMENT EDITOR
GEORGE J. HADEMENOS

SCHAUM'S OUTLINE SERIES
McGRAW-HILL

New York   Chicago   San Francisco   Lisbon   London   Madrid
Mexico City   Milan   New Delhi   San Juan
Seoul   Singapore   Sydney   Toronto

FRANK AYRES, JR., Ph.D. was formerly professor and head of the Department of Mathematics at Dickinson College in Carlisle, Pennsylvania. He is the author of eight *Schaum's Outlines*, including *Calculus, Differential Equations,* and *College Mathematics.*

ROBERT E. MOYER teaches mathematics at Fort Valley State University in Fort Valley, Georgia and was formerly head of the Department of Mathematics and Physics. He was previously the mathematics consultant for a five-county public school cooperative and taught high school mathematics in Illinois. He received his Ph.D. in Mathematics Education from the University of Illinois and his B.S. and M.S. degrees from Southern Illinois University.

GEORGE J. HADEMENOS has taught at the University of Dallas and done research at the University of Massachusetts Medical Center and the University of California at Los Angeles. He holds a B.S. degree from Angelo State University and both M.S. and Ph.D. degrees from the University of Texas at Dallas. He is the author of several books in the Schaum's Outline and Schaum's Easy Outline series.

1 2 3 4 5 6 7 8 9 0 DOC DOC 0 9 8 7 6 5 4 3 2

ISBN 0-07-138318-2

McGraw-Hill

A Division of The McGraw-Hill Companies

Contents

Chapter 1
ANGLES AND APPLICATIONS

Introduction

Trigonometry, as the name implies, is concerned with the measurement of the parts, sides, and angles of a triangle. Plane trigonometry, which is the topic of this book, is restricted to triangles lying in a plane. Trigonometry is based on certain ratios, called trigonometric functions, to be defined in the next chapter. The early applications of the trigonometric functions were to surveying, navigation, and engineering. These functions also play an important

role in the study of all sorts of vibratory phenomena—sound, light, electricity, etc. As a consequence, a considerable portion of the subject matter is concerned with a study of properties of and relations among the trigonometric functions.

Plane Angle

The plane angle *XOP*, Figure 1-1, is formed by the two rays *OX* and *OP*. The point *O* is called the *vertex* and the half lines are called the *sides* of the angle.

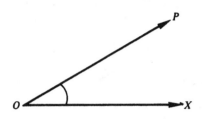

Figure 1-1

An angle is called *positive* if the direction of rotation (indicated by a curved arrow) is counterclockwise and *negative* if the direction of rotation is clockwise. The angle is positive in Figure 1-2 (a) and (c), negative in Figure 1-2 (b).

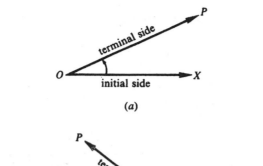

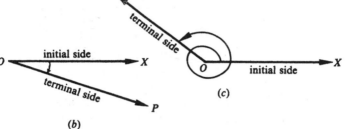

Figure 1-2

Measures of Angles

A *degree* (°) is defined as the measure of the central angle subtended by an arc of a circle equal to 1/360 of the circumference of the circle.

A *minute* (′) is 1/60 of a degree; a second (″) is 1/60 of a minute, or 1/3600 of a degree.

When changing angles in decimals to minutes and seconds, the general rule is that angles in tenths will be changed to the nearest minute and all other angles will be rounded to the nearest hundredth and then changed to the nearest second. When changing angles in minutes and seconds to decimals, the results in minutes are rounded to tenths, and angles in seconds have the results rounded to hundredths.

Example 1.1

(a) $62.4° = 62° + 0.4 \,(60′) = 62°24′$

(b) $23.9° = 23° + 0.9\ (60') = 23°54'$

(c) $29.23° = 29° + 0.23\ (60') = 29°13.8' = 29°13' + 0.8\ (60'')$
     $= 29°13'48''$

(d) $37.47° = 37° + 0.47\ (60') = 37°28.2' = 37°28' + 0.2\ (60'')$
     $= 37°28'12''$

(e) $78°17' = 78° + 17°/60 = 78.3°$ (rounded to tenths)

(f) $58°22'16'' = 58° + 22°/60 + 16°/3600 = 58.37°$
     (rounded to hundredths)

A *radian* (rad) is defined as the measure of the central angle subtended by an arc of a circle equal to the radius of the circle. (See Figure 1-3.)

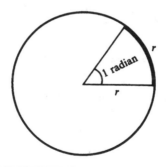

Figure 1-3

The circumference of a circle $= 2\pi$ (radius) and subtends an angle of 360°; therefore

$$1 \text{ radian} = 180°/\pi = 57.296° = 57°17'45''$$

and
$$1 \text{ degree} = \pi/180 \text{ radian} = 0.017453 \text{ rad}$$
where $\pi = 3.14159$.

Example 1.2   (a) $7/12\ \pi$ rad $= (7\pi/12)(180°/\pi) = 105°$

(b) $50° = 50(\pi/180)$ rad $= (5\pi/18)$ rad

Arc Length

On a circle of radius r, a central angle of θ radians, Figure 1-4, intercepts an arc of length

$$s = r\theta$$

that is, arc length = radius × central angle in radians.

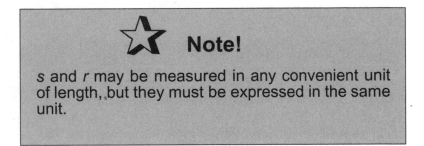

Note!

s and r may be measured in any convenient unit of length, but they must be expressed in the same unit.

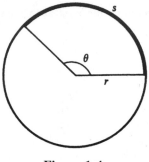

Figure 1-4

Example 1.3   (a) On a circle of radius 30 in, the length of the arc intercepted by a central angle of 1/3 rad is:

$$s = r\theta = 30(1/3) = 10 \text{ in}$$

(b) On the same circle, a central angle of 50° intercepts an arc of length:

$$s = r\theta = 30(5\pi/18) = 25\pi/3 \text{ in}$$

Lengths of Arcs on a Unit Circle

The correspondence between points on a real number line and the points on a unit circle, $x^2 + y^2 = 1$, with its center at the origin is shown in Figure 1-5.

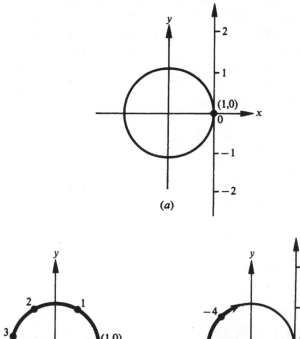

(a)

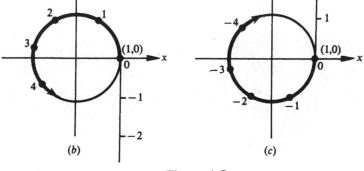

(b)

(c)

Figure 1-5

The zero, 0, on the number line is matched with the point (1, 0) as shown in Figure 1-5(a). The positive real numbers are wrapped around the circle in a counterclockwise direction, Figure 1-5(b), and the negative real numbers are wrapped around the circle in a clockwise direction, Figure 1-5(c). Every point on the unit circle is matched with many real numbers, both positive and negative.

The radius of a unit circle has length 1. Hence, the circumference of the circle, given by $2\pi r$, is 2π. The distance halfway around is π and $\frac{1}{4}$ the way around is $\pi/2$. Each positive number is paired with a length of an arc s, and since $s = r\theta = 1 \cdot \theta = \theta$, each real number is paired with an angle θ in radian measure. Likewise, each negative real number is paired with the negative of the length of an arc and, therefore, with a negative angle in radian measure.

Area of a Sector

The area K of a sector of a circle, with radius r and central angle θ radians is

$$K = \frac{1}{2} r^2 \theta$$

that is, the area of a sector = $\frac{1}{2} \times$ the radius $\times$ the radius $\times$ the central angle in radians.

⭐ **Note!**

K will be measured in the square unit of area that corresponds to the length unit used to measure *r*.

Example 1.4   For a circle of radius 18 cm, the area of a sector intercepted by a central angle of 50° is

$$K = \frac{1}{2} r^2 \theta = \frac{1}{2} (18)^2 \, 5\pi/18 = 45 \, \pi \text{ cm}^2 \text{ or } 141 \text{ cm}^2 \text{ (rounded)}$$

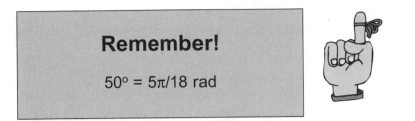

Remember!

$50° = 5\pi/18$ rad

Angular Velocity

The relationship between the linear velocity v and the angular velocity ω (the Greek letter *omega*) for an object with radius r is

$$v = r\omega$$

where ω is measured in radians per unit of time and v is distance per unit of time.

> v and ω use the same unit of time
> and r and v use the same linear unit

Example 1.5 A bicycle with 20-in wheels is traveling down a road at 15 mi/h. Find the angular velocity of the wheel in revolutions per minute.

Because the radius is 10 in and the angular velocity is to be in revolutions per minute (r/min), change the linear velocity 15 mi/h to units of in/min.

$v = 15$ mi/h
$= (15/1)(\text{mi/h})\cdot(5280/1)(\text{ft/mi})\cdot(12/1)(\text{in/ft})\cdot(1/60)(\text{h/min})$
$= 15,840$ in/min

$\omega = v/r = (15,840/10)(\text{rad/min}) = 1584$ rad/min

To change ω to r/min, we multiply by ½ π revolution per radian (r/rad):

$\omega = 1584$ rad/min $= (1584/1)(\text{rad/min})\cdot(1/2\pi)(\text{r/rad})$
$= (792/\pi)(\text{r/min})$ or 252 r/min

Solved Problem 1.1 Assuming the Earth to be a sphere of radius 3960 mi, find the distance of a point 36°N latitude from the equator.

Solution. Since $36° = \pi/5$ rad, $s = r\theta = 3960(\pi/5) = 2488$ mi.

Solved Problem 1.2 Two cities 270 mi apart lie on the same meridian. Find their difference in latitude.

Solution.

$$\theta = \frac{s}{r} = \frac{270}{3960} = \frac{3}{44}\text{rad} \quad \text{or} \quad 3°54.4'$$

Chapter 2
TRIGONOMETRIC FUNCTIONS OF A GENERAL ANGLE

Coordinates on a Line

A *directed line* is a line on which one direction is taken as positive and the other as negative. The positive direction is indicated by an arrowhead.

11

A *number scale* is established on a directed line by choosing a point O (Figure 2-1) called the *origin* and a unit of measure $OA = 1$.

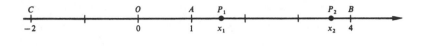

Figure 2-1

On this scale, B is 4 units to the right of O (that is, in the positive direction from O) and C is 2 units to the left of O (that is, in the negative direction from O). The directed distance $OB = +4$ and the directed distance $OC = -2$. It is important to note that since the line is directed, $OB \neq BO$ and $OC \neq CO$. The directed distance $BO = -4$, is measured contrary to the indicated positive direction, and the directed distance $CO = +2$. Then, $CB = CO + OB = 2 + 4 = 6$ and $BC = BO + OC = -4 + (-2) = -6$.

Coordinates in a Plane

A *rectangular coordinate system* in a plane consists of two number scales (called axes), one horizontal and the other vertical, whose point of intersection (*origin*) is the origin on each scale. It is customary to choose the positive direction on each axis as indicated in the figure, that is, positive to the right on the horizontal axis or x axis and positive upward on the vertical or y axis. For convenience, we shall assume the same unit of measure on each axis.

By means of such a system, the position of any point P in the plane is given by its (directed) distances, called *coordinates*, from the axes. The x coordinate or *abscissa* of a point P (Figure 2-2) is the directed distance $BP = OA$ and the y coordinate or *ordinate* is the directed distance $AP = OB$. A point P with abscissa x and ordinate y will be denoted by $P(x, y)$.

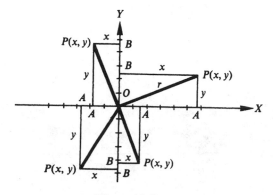

Figure 2-2

The axes divide the plane into four parts, called *quadrants*, which are numbered in a counterclockwise direction I, II, III, and IV. The numbered quadrants, together with the signs of the coordinates of a point in each, are shown in Figure 2-3.

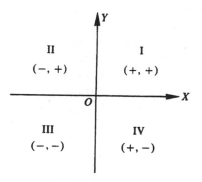

Figure 2-3

The undirected distance r of any point $P(x, y)$ from the origin, called the *distance of P* or the hypotenuse (sometimes known as the *radius vector of P*), is given by

$$r = \sqrt{x^2 + y^2}$$

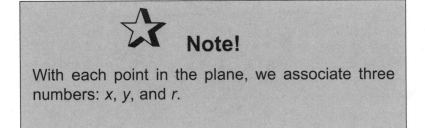

Note!

With each point in the plane, we associate three numbers: x, y, and r.

Angles in Standard Position

With respect to a rectangular coordinate system, an angle is said to be *in standard position* when its vector is at the origin and its initial side coincides with the positive x axis.

An angle is said to be a *first-quadrant angle* or to be *in the first quadrant* if, when in standard position, its terminal side falls in that quadrant. Similar definitions hold for the other quadrants. For example, the angles 30°, 59°, and −330° are first-quadrant angles (Figure 2-4(*a*)); 119° is a second-quadrant angle; −119° is a third-quadrant angle; and −10° and 710° are fourth-quadrant angles (Figure 2-4(*b*)).

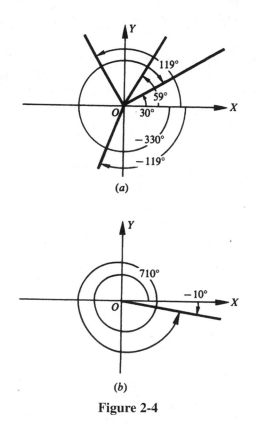

Figure 2-4

Two angles which, when placed in standard position, have coincident terminal sides are called *coterminal angles*. For example, 30° and −330°, and −10° and 710° are pairs of coterminal angles. There are an unlimited number of angles coterminal with a given angle. Coterminal angles for any given angle can be found by adding integer multiples of 360° to the degree measure of the given angle.

The angles 0°, 90°, 180°, and 270° and all the angles coterminal with them are called *quadrantal angles*.

Trigonometric Functions of a General Angle

Let θ be an angle (not quadrantal) in standard position and let $P(x, y)$ be any point, distinct from the origin, on the terminal side of the angle. The six trigonometric functions of θ are defined, in terms of the abscissa, ordinate, and distance of P, as follows:

$$\text{sine } \theta = \sin \theta = \frac{ordinate}{hypotenuse} = \frac{y}{r}$$

$$\text{cosine } \theta = \cos \theta = \frac{abscissa}{hypotenuse} = \frac{x}{r}$$

$$\text{tangent } \theta = \tan \theta = \frac{ordinate}{abscissa} = \frac{y}{x}$$

$$\text{cotangent } \theta = \cot \theta = \frac{abscissa}{ordinate} = \frac{x}{y}$$

$$\text{secant } \theta = \sec \theta = \frac{hypotenuse}{abscissa} = \frac{r}{x}$$

$$\text{cosecant } \theta = \csc \theta = \frac{hypotenuse}{ordinate} = \frac{r}{y}$$

As an immediate consequence of these definitions, we have the so-called reciprocal relations:

$\sin \theta = 1/\csc \theta$ $\qquad\qquad$ $\tan \theta = 1/\cot \theta$ $\qquad\qquad$ $\sec \theta = 1/\cos \theta$

$\cos \theta = 1/\sec \theta$ $\qquad\qquad$ $\cot \theta = 1/\tan \theta$ $\qquad\qquad$ $\csc \theta = 1/\sin \theta$

Because of these reciprocal relationships, one function in each pair of reciprocal trigonometric functions has been used more frequently than the other. The more frequently used trigonometric functions are sine, cosine, and tangent.

It is evident from the diagrams in Figure 2-5 that the values of the trigonometric functions of θ change as θ changes.

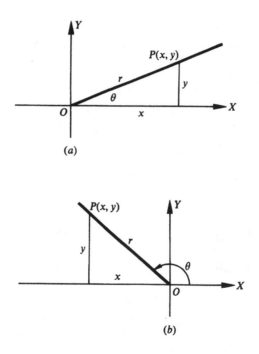

(a)

(b)

Figure 2-5

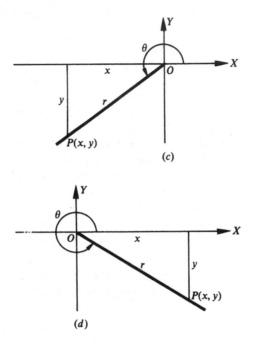

(c)

(d)

Figure 2-5, cont.

Quadrant Signs of the Functions

Since r is always positive, the signs of the functions in the various quadrants depend on the signs of x and y. To determine these signs, one may visualize the angle in standard position or use some device as shown in Figure 2-6 in which only the functions having positive signs are listed.

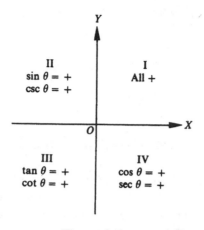

Figure 2-6

When an angle is given, its trigonometric functions are uniquely determined. When, however, the value of one function of an angle is given, the angle is not uniquely determined. For example, if $sin\ \theta = \frac{1}{2}$, then $\theta = 30°$, $150°$, $390°$, $510°$, ... In general, two possible positions of the terminal side are found, for example, the terminal sides of $30°$ and $150°$ in the above illustration. The exceptions to this rule occur when the angle is quadrantal.

Trigonometric Functions of Quadrantal Angles

For a quadrantal angle, the terminal side coincides with one of the axes. A point P, distinct from the origin, on the terminal side has either $x = 0$ and $y \neq 0$, or $x \neq 0$ and $y = 0$. In either case, two of the six functions will not be defined. For example, the terminal side of the angle $0°$ coincides with the positive x axis and the ordinate of P is 0. Since the ordinate occurs in the denominator of the ratio defining the cotangent and cosecant, these functions are not defined. In this book, *undefined* will be used instead of a numerical value in such cases, but some authors indicate this by writing $cot\ 0° = \infty$ and others write $cot\ 0° = \pm\infty$.

Angle θ	$\sin \theta$	$\cos \theta$	$\tan \theta$	$\cot \theta$	$\sec \theta$	$\csc \theta$
0°	0	1	0	Undefined	1	Undefined
90°	1	0	Undefined	0	Undefined	1
180°	0	−1	0	Undefined	−1	Undefined
270°	−1	0	Undefined	0	Undefined	−1

Undefined Trigonometric Functions

It has been noted that cot 0° and csc 0° are not defined since division by zero is never allowed, but the values of these functions for angles near 0° are of interest. In Figure 2-7(a), take θ to be a small positive angle in standard position and, on its terminal side, take $P(x, y)$ to be at a distance r from O. Now x is slightly less than r and y is positive and very small; then cot $\theta = x/y$ and csc $\theta = r/y$ are positive and very large. Next, let θ decrease toward 0° with P remaining at a distance r from O. Now, x increases but is always less than r while y decreases but remains greater than 0; thus cot θ and csc θ become larger and larger. (To see this, take $r = 1$ and compute csc θ when $y = 0.1, 0.01, 0.001, \ldots$.) This state of affairs is indicated by "If θ approaches $0°^+$, then cot θ approaches $+\infty$," which is what is meant when writing cot 0° = $+\infty$.

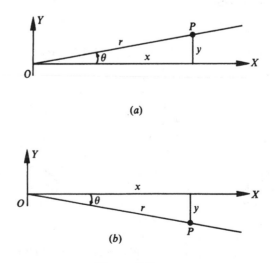

(a)

(b)

Figure 2-7

Next suppose, as in Figure 2-7(b), that θ is a negative angle but close to $0°$ and take $P(x,y)$ on its terminal side at a distance r from O. Then x is positive and slightly smaller than r while y is negative and has a small absolute value. Both cot θ and csc θ are negative with large absolute values. Next, let θ increase toward $0°$ with P remaining at a distance r from O. Now x increases but is always less than r, while y remains negative with an absolute value decreasing toward 0; thus cot θ and csc θ remain negative but have absolute values that get larger and larger. This situation is indicated by "If θ approaches $0°^-$, then cot θ approaches $-\infty$," which is what is meant when writing cot $0° = -\infty$.

In each of these cases, cot $0° = +\infty$ and cot $0° = -\infty$, the use of the = sign does not have the standard meaning of "equals" and should be used with caution since cot $0°$ is undefined and ∞ is not a number. The notation is used as a short way to describe a special situation for trigonometric functions.

The behavior of other trigonometric functions that become undefined can be explored in a similar manner. The following chart summarizes the behavior of each trigonometric function that becomes undefined for angles from $0°$ up to $360°$.

Angle θ	Function Values
$\theta \rightarrow 0°^+$	cot $\theta \rightarrow +\infty$ and csc $\theta \rightarrow +\infty$
$\theta \rightarrow 0°^-$	cot $\theta \rightarrow -\infty$ and csc $\theta \rightarrow -\infty$
$\theta \rightarrow 90°^-$	tan $\theta \rightarrow +\infty$ and sec $\theta \rightarrow +\infty$
$\theta \rightarrow 90°^+$	tan $\theta \rightarrow -\infty$ and sec $\theta \rightarrow -\infty$
$\theta \rightarrow 180°^-$	cot $\theta \rightarrow -\infty$ and csc $\theta \rightarrow +\infty$
$\theta \rightarrow 180°^+$	cot $\theta \rightarrow +\infty$ and csc $\theta \rightarrow -\infty$
$\theta \rightarrow 270°^-$	tan $\theta \rightarrow +\infty$ and sec $\theta \rightarrow -\infty$
$\theta \rightarrow 270°^+$	tan $\theta \rightarrow -\infty$ and sec $\theta \rightarrow +\infty$

The + means the value is greater than the number stated; $180°^+$ means values greater than $180°$. The minus sign means the value is less than the number stated; $90°^-$ means values less than $90°$.

Coordinates of Points on a Unit Circle

Let s be the length of an arc on a unit circle $x^2 + y^2 = 1$; each s is paired with an angle θ in radians. Using the point $(1, 0)$ as the initial point of the arc and $P(x, y)$ as the terminal point of the arc, as in Figure 2-8, we can determine the coordinates of P in terms of the real number s.

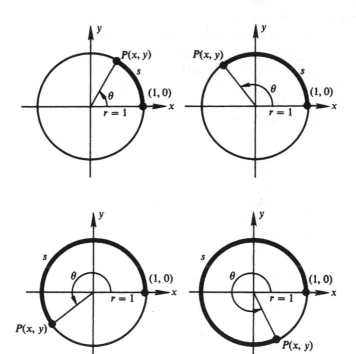

Figure 2-8

For any angle θ, $\cos \theta = x/r$ and $\sin \theta = y/r$. On a unit circle, $r = 1$ and the arc length $s = r\theta = \theta$ and $\cos \theta = \cos s = x/1 = x$ and $\sin \theta = \sin s = y/1 = y$. The point P associated with the arc length s is determined by $P(x, y) = P(\cos s, \sin s)$. The wrapping function W maps real numbers s onto points P of the unit circle denoted by

$$W(s) = (\cos s, \sin s)$$

Some arc lengths are paired with points on the unit circle whose coordinates are easily determined. If $s = 0$, the point is $(1, 0)$; for $s = \pi/2$, one-fourth the way around the unit circle, the point is $(0, 1)$; $s = \pi$ is paired with $(-1, 0)$; and $s = 3\pi/2$ is paired with $(0, -1)$. These values are summarized in the following chart.

s	$P(x, y)$	$\cos s$	$\sin s$
0	$(1, 0)$	1	0
$\pi/2$	$(0, 1)$	0	1
π	$(-1, 0)$	-1	0
$3\pi/2$	$(0, -1)$	0	-1

Circular Functions

Each arc length s determines a single ordered pair $(\cos s, \sin s)$ on a unit circle. Both s and $\cos s$ are real numbers and define a function $(s, \cos s)$ which is called the *circular function cosine*.
Likewise, s and $\sin s$ are real numbers and define a function $(s, \sin s)$ which is called the *circular function sine*. These functions are called *circular functions* since both $\cos s$ and $\sin s$ are coordinates on a unit circle. The cir- cular functions $\sin s$ and $\cos s$ are similar to the trigonometric functions $\sin \theta$ and $\cos \theta$ in all regards since any angle in degree measure can be converted to radian measure and this radian-measure angle is paired with an arc s on the unit circle. The important distinction for circular functions is that since $(s, \cos s)$ and $(s, \sin s)$ are ordered pairs of real numbers, all properties and procedures for functions of real numbers apply to circular functions.

The remaining circular functions are defined in terms of $\cos s$ and $\sin s$.

$$\tan s = \frac{\sin s}{\cos s} \qquad \text{for } s \neq \frac{\pi}{2} + k\pi \text{ where } k \text{ is an integer}$$

$$\cot s = \frac{\cos s}{\sin s} \qquad \text{for } s \neq k\pi \text{ where } k \text{ is an integer}$$

$$\sec s = \frac{1}{\cos s} \qquad \text{for } s \neq \frac{\pi}{2} + k\pi \text{ where } k \text{ is an integer}$$

$$\csc s = \frac{1}{\sin s} \qquad \text{for } s \neq k\pi \text{ where } k \text{ is an integer}$$

It should be noted that the circular functions are defined everywhere that the trigonometric functions are defined and that the values left out of the domains correspond to values where the trigonometric functions are undefined.

In any application, there is no need to distinguish between trigonometric functions of angles in radian measure and circular functions of real numbers.

Solved Problem 2.1 Using a rectangular coordinate system, locate the following points and find the value of r for each: $A(1, 2)$; $B(-3, 4)$; $C(-3, -3\sqrt{3}\,)$; $D(4, -5)$ as shown in Figure 2-9.

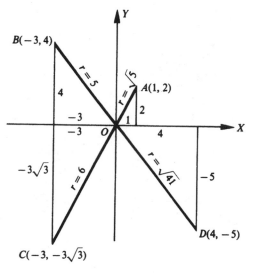

Figure 2-9

Solution.

For A: $r = \sqrt{x^2 + y^2} = \sqrt{1+4} = \sqrt{5}$

For B: $r = \sqrt{x^2 + y^2} = \sqrt{9+16} = 5$

For C: $r = \sqrt{x^2 + y^2} = \sqrt{9+27} = 6$

For D: $r = \sqrt{x^2 + y^2} = \sqrt{16+25} = \sqrt{41}$

Solved Problem 2.2 Construct the following angles in standard position and determine those which are coterminal: $125°$, $210°$, $-150°$, $385°$, $930°$, $-370°$, $-955°$, and $-870°$, as shown in Figure 2-10.

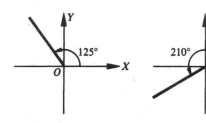

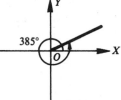

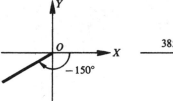

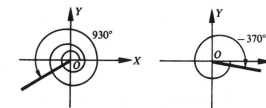

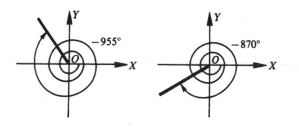

Figure 2-10

Solution. The angles in standard position are shown in Figure 2-10. The angles 125° and −955° are coterminal since −955° = 125° − 3 · 360° (or since $125° = -955° + 3 \cdot 360°$). The angles $210°$, $-150°$, $930°$, and $-870°$ are coterminal since $-150° = 210° - 1 \cdot 360°$, $930° = 210° + 2 \cdot 360°$, and $-870° = 210° - 3 \cdot 360°$. From Figure 2-10, it can be seen that there is only one first-quadrant angle, $385°$, and only one-fourth-quadrant angle, $-370°$, so these angles cannot be coterminal with any of the other angles.

Solved Problem 2-3 In what quadrants may θ terminate, if: (a) sin θ is positive?; (b) cos θ is negative?; (c) tan θ is negative?; (d) sec θ is positive?

Solution.

(a) Since sin θ is positive, y is positive. Then x may be positive or negative and θ is a first- or second-quadrant angle.

(b) Since cos θ is negative, x is negative. Then y may be positive or negative and θ is a second- or third-quadrant angle.

(c) Since tan θ is negative, either y is positive or y is negative and x is positive. Thus, θ may be a second- or fourth-quadrant angle.

(d) Since sec θ is positive, x is positive. Thus, θ may be a first- or fourth-quadant angle.

Chapter 3
TRIGONOMETRIC FUNCTIONS OF AN ACUTE ANGLE

IN THIS CHAPTER:

Trigonometric Functions of an Acute Angle

In dealing with any right triangle, it will be convenient (Figure 3-1) to denote the vertices as A, B, and C with C the vertex of the right angle, to denote the angles of the triangles as A, B, and C with $C = 90°$, and to denote the sides opposite the angles as a, b, and c, respectively. With respect to  angle A, a will be called the *opposite side* and b will be called the *adjacent side*; with respect to angle B, b will be called the *opposite side* and a the *adjacent side*. Side c will always be called the *hypotenuse*.

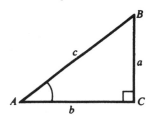

Figure 3-1

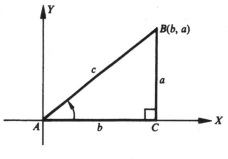

Figure 3-2

If now the right triangle is placed in a coordinate system (Figure 3-2) so angle A is in standard position, the point B on the terminal side

of angle A has coordinates (b, a), and the distance $c = \sqrt{a^2 + b^2}$.
Then the trigonometric functions of angle A may be defined in terms of
the sides of the right triangle, as follows.

$$\sin A = \frac{opposite\ side}{hypotenuse} = \frac{a}{c} \qquad \cos A = \frac{adjacent\ side}{hypotenuse} = \frac{b}{c}$$

$$\tan A = \frac{opposite\ side}{adjacent\ side} = \frac{a}{b} \qquad \cot A = \frac{adjacent\ side}{opposite\ side} = \frac{b}{a}$$

$$\sec A = \frac{hypotenuse}{adjacent\ side} = \frac{c}{b} \qquad \csc A = \frac{hypotenuse}{opposite\ side} = \frac{c}{a}$$

Trigonometric Functions of Complementary Angles

The acute angles A and B of the right triangle ABC are complementary;
that is, $A + B = 90°$. From Figure 3-1, we have

$\sin B = b/c = \cos A \qquad\qquad \cos B = a/c = \sin A$

$\tan B = b/a = \cot A \qquad\qquad \cot B = a/b = \tan A$

$\sec B = c/a = \csc A \qquad\qquad \csc B = c/b = \sec A$

These relations associate the functions in pairs—sine and cosine, tangent and cotangent, secant and cosecant—each function of a pair being
called the *cofunction* of the other. Thus, any function of an acute angle
is equal to the corresponding cofunction of the complementary angle.

Example 3.1 Find the values of the trigonometric functions of the angles of the right triangle ABC in Figure 3-3.

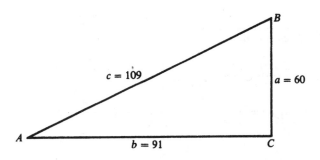

Figure 3-3

$$\sin A = \frac{opposite\ side}{hypotenuse} = \frac{a}{c} = \frac{60}{109}$$

$$\cos A = \frac{adjacent\ side}{hypotenuse} = \frac{b}{c} = \frac{91}{109}$$

$$\tan A = \frac{opposite\ side}{adjacent\ side} = \frac{a}{b} = \frac{60}{91}$$

$$\cot A = \frac{adjacent\ side}{opposite\ side} = \frac{b}{a} = \frac{91}{60}$$

$$\sec A = \frac{hypotenuse}{adjacent\ side} = \frac{c}{b} = \frac{109}{91}$$

$$\csc A = \frac{hypotenuse}{opposite\ side} = \frac{c}{a} = \frac{109}{60}$$

$$\sin B = \frac{opposite\ side}{hypotenuse} = \frac{b}{c} = \frac{91}{109}$$

$$\cos B = \frac{adjacent\ side}{hypotenuse} = \frac{a}{c} = \frac{60}{109}$$

$$\tan B = \frac{opposite\ side}{adjacent\ side} = \frac{b}{a} = \frac{91}{60}$$

$$\cot B = \frac{adjacent\ side}{opposite\ side} = \frac{a}{b} = \frac{60}{91}$$

$$\sec B = \frac{hypotenuse}{adjacent\ side} = \frac{c}{a} = \frac{109}{60}$$

$$\csc B = \frac{hypotenuse}{opposite\ side} = \frac{c}{b} = \frac{109}{91}$$

Trigonometric Functions of 30°, 45°, and 60°

The special acute angles 30°, 45°, and 60° have trigonometric function values that can be computed exactly. For each fraction that had an irrational number denominator, only the equivalent fraction with a rational number denominator is stated in the table.

Angle θ	$\sin \theta$	$\cos \theta$	$\tan \theta$	$\cot \theta$	$\sec \theta$	$\csc \theta$
30°	$\frac{1}{2}$	$\frac{1}{2}\sqrt{3}$	$\frac{1}{3}\sqrt{3}$	$\sqrt{3}$	$\frac{2}{3}\sqrt{3}$	2
45°	$\frac{1}{2}\sqrt{2}$	$\frac{1}{2}\sqrt{2}$	1	1	$\sqrt{2}$	$\sqrt{2}$
60°	$\frac{1}{2}\sqrt{3}$	$\frac{1}{2}$	$\sqrt{3}$	$\frac{1}{3}\sqrt{3}$	2	$\frac{2}{3}\sqrt{3}$

Trigonometric Function Values

For many application problems, values of trigonometric functions are needed for angles that are not special angles. These values may be found in tables of trigonometric functions or by using a scientific calculator. A two-decimal-place table of trigonometric function values is shown below.

Angle θ	$\sin \theta$	$\cos \theta$	$\tan \theta$	$\cot \theta$	$\sec \theta$	$\csc \theta$
15°	0.26	0.97	0.27	3.73	1.04	3.86
20°	0.34	0.94	0.36	2.75	1.06	2.92
30°	0.50	0.87	0.58	1.73	1.15	2.00
40°	0.64	0.77	0.84	1.19	1.31	1.56
45°	0.71	0.71	1.00	1.00	1.41	1.41
50°	0.77	0.64	1.19	0.84	1.56	1.31
60°	0.87	0.50	1.73	0.58	2.00	1.15
70°	0.94	0.34	2.75	0.36	2.92	1.06
75°	0.97	0.26	3.73	0.27	3.86	1.04

When using a calculator to find values for trigonometric functions, be sure to follow the procedure indicated in the instruction manual for your calculator. In general, the procedure is (1) make sure the calculator is in degree mode, (2) enter the number of degrees in the angle, (3) press the key for the trigonometric function wanted, and (4) read the function value from the display.

Example 3.3 Find tan 15° using a calculator.

Solution. With the calculator in degree mode, enter 15 and press the (tan) key. The number 0.267949 will appear on the display; thus tan 15° = 0.267949. The number of digits that are displayed depends on the calculator used, but most scientific calculators show at least six digits.

Using a calculator to find an acute angle when the value of a trigonometric function is given, requires the use of the inverse (inv) key or the second function (2d) key. The value of the function is entered, the (inv) key is pressed, and then the trigonometric function key is pressed. The degree mode is used to get answers in degree measure.

Accuracy of Results Using Approximations

When using approximate numbers, the results need to be rounded. In this chapter, we will report angles to the nearest degree and lengths to the nearest unit. If a problem has intermediate values to be computed, wait to round numbers until the final result is found. Each intermediate value should have at least one more digit than the final result is to have so that each rounding does not directly involve the unit of accuracy.

Selecting the Function in Problem Solving

In finding a side of a right triangle when an angle and a side are known, there are two trigonometric functions that can be used, a function and its reciprocal. When manually solving the problem, the choice is usually made so the unknown side is in the numera-tor of the fraction. This is done so the operation needed to solve the equation will be multiplication rather than division. When a calculator is used, the function selected is sine, cosine, or tangent since these functions are represented by keys on the calculator.

Example 3.3 A support wire is anchored 12 m up from the base of a flagpole and the wire makes a 15° angle with the ground, as shown in Figure 3-4. How long is the wire?

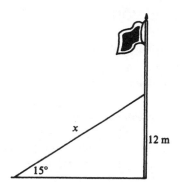

Figure 3-4

Solution. From Figure 3-4, it can be seen that both sin 15° and csc 15° angle involve the known length 12 m and the requested length x. Either function can be used to solve the problem. The manual solution, that is, using tables and not a calculator, is easier using csc 15°, but not all trigonometric tables list values for secant and cosecant. The calculator solution will use sin 15° since there is no function key for cosecant.

	Manual Solution		Calculator Solution

$$\csc 15° = \frac{x}{12} \qquad \text{or} \qquad \sin 15° = \frac{12}{x} \qquad\qquad \sin 15° = \frac{12}{x}$$

$$x = 12 \csc 15° \qquad\qquad x = \frac{12}{\sin 15°} \qquad\qquad x = \frac{12}{\sin 15°}$$

$$x = 12(3.86) \qquad\qquad x = \frac{12}{0.26} \qquad\qquad x = \frac{12}{0.258819}$$

$$x = 46.32 \qquad\qquad x = 46.15 \qquad\qquad x = 46.3644$$

$$x = 46 \text{ m} \qquad\qquad x = 46 \text{ m} \qquad\qquad x = 46 \text{ m}$$

The wire is 46 m long.

Angles of Depression and Elevation

An angle of depression is the angle from the horizontal down to the line of sight from the observer to an object below. The angle of elevation is the angle from the horizontal up to the line of sight from the observer to an object above.

In Figure 3-5, the angle of depression from point A to point B is α and the angle of elevation from point B to point A is β. Since both angles are measured from horizontal lines, which are parallel, the line of sight AB is a transversal, and since alternate interior angles for parallel lines are equal, $\alpha = \beta$.

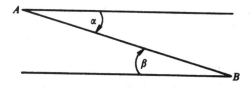

Figure 3-5

Solved Problem 3.1 Find the trigonometric functions of the acute angles of the right triangle ABC, Figure 3-6, given $b = 24$ and $c = 25$.

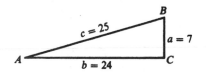

Figure 3-6

Solution. Since $a^2 = c^2 - b^2 = (25)^2 - (24)^2 = 49$, $a = 7$. Then,

$$\sin A = \frac{opposite\ side}{hypotenuse} = \frac{a}{c} = \frac{7}{25}$$

$$\cos A = \frac{adjacent\ side}{hypotenuse} = \frac{b}{c} = \frac{24}{25}$$

$$\tan A = \frac{opposite\ side}{adjacent\ side} = \frac{a}{b} = \frac{7}{24}$$

$$\cot A = \frac{adjacent\ side}{opposite\ side} = \frac{b}{a} = \frac{24}{7}$$

$$\sec A = \frac{hypotenuse}{adjacent\ side} = \frac{c}{b} = \frac{25}{24}$$

$$\csc A = \frac{hypotenuse}{opposite\ side} = \frac{c}{a} = \frac{25}{7}$$

and

$\sin B = 24/25$ $\cos B = 7/25$ $\tan B = 24/7$

$\csc B = 25/24$ $\sec B = 25/7$ $\cot B = 7/24$

Solved Problem 3.2 A tree 100 ft tall casts a shadow 120 ft long, as shown in Figure 3-7. Find the angle of elevation of the sun.

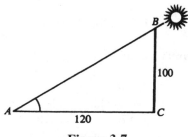

Figure 3-7

Solution. From Figure 3-7, $CB = 100$, $AC = 120$, and we want to find A.

$$\tan A = \frac{CB}{AC} = \frac{100}{120} = 0.83$$

$$A = 40°$$

Solved Problem 3.3 From the top of a lighthouse, 120 m above the sea, the angle of depression of a boat is 15°, as shown in Figure 3-8. How far is the boat from the lighthouse?

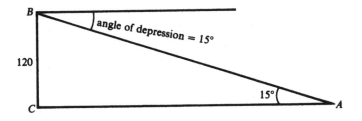

Figure 3-8

Solution. In Figure 3-8, the right triangle ABC has $A = 15°$ and $CB = 120$. Then the solution follows.

$$\cot A = AC/CB$$

$$AC = CB \cot A = 120 \cot 15° = 120(3.73) = 447.6$$

The boat is 448 m from the lighthouse.

Solved Problem 3.4 A tower standing on level ground is due north of point A and due west of point B, a distance c ft from A. If the angles of elevation of the top of the tower as measured from A and B are α and β, respectively, find the height h of the tower, as shown in Figure 3-9.

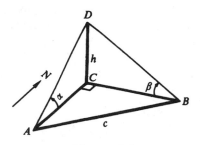

Figure 3-9

Solution. In the right triangle ACD of Figure 3-9, cot $\alpha = AC/h$; and in the right triangle BCD, cot $\beta = BC/h$. Then $AC = h$ cot α and $BC = h$ cot β.

Since ABC is a right triangle, $(AC)^2 + (BC)^2 = c^2 = h^2 (\cot \alpha)^2$ and $BC = h^2 (\cot \beta)^2$ and

$$h = \frac{c}{\sqrt{(\cot \alpha)^2 + (\cot \beta)^2}}$$

Solved Problem 3.5 If holes are to be spaced regularly on a circle, show that the distance d between the centers of two successive holes is given by $d = 2r \sin (180°/n)$, where $r =$ the radius of the circle and $n =$ the number of holes. Find d when $r = 20$ in and $n = 4$.

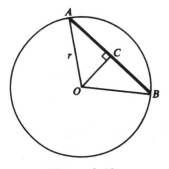

Figure 3-10

Solution. In Figure 3-10, let A and B be the centers of two consecutive holes on the circle of radius r and center O. Let the bisector of the angle O of the triangle AOB meet AB at C. In right triangle AOC,

$$\sin \angle AOC = \frac{AC}{r} = \frac{d/2}{r} = \frac{d}{2r}$$

Then,

$$D = 2r \sin \angle AOC$$

$$= 2r \sin \frac{1}{2} \angle AOB$$

$$= 2r \sin \frac{1}{2} \left(\frac{360^{\,o}}{n} \right)$$

$$= 2r \sin \frac{180^{\,o}}{n}$$

When $r = 20$ and $n = 4$, $d = 2 \cdot 20 \sin 45° = 2 \cdot 20 \left(\sqrt{2}/2 \right) = 20\sqrt{2}$

Chapter 4
PRACTICAL
APPLICATIONS

IN THIS CHAPTER:

✔ *Bearing*
✔ *Vectors*
✔ *Vector Addition*
✔ *Components of a Vector*
✔ *Air Navigation*
✔ *Inclined Plane*

Bearing

The bearing of a point *B* from a point *A*, in a horizontal plane, is usually defined as the angle (always acute) made by the ray drawn from *A* through *B* with the north-south line through *A*. The bearing is then read from the north or south line toward the east or west. The angle used in expressing a bearing is usually stated in degrees and minutes. For example, see Figure 4-1.

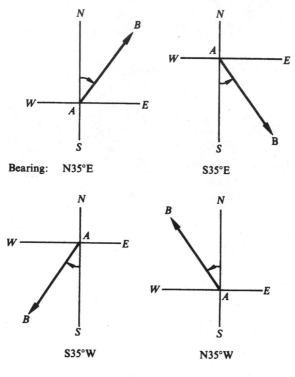

Figure 4-1

In aeronautics, the bearing of *B* from *A* is more often given as the angle made by the ray *AB* with the north line through *A*, measured clockwise from the north (i.e., from the north around through the east). For example, see Figure 4-2.

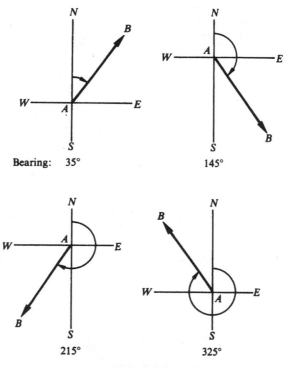

Figure 4-2

Vectors

Any physical quantity, like force or velocity, that has both magnitude and direction is called a *vector quantity*. A vector quantity may be represented by a directed line segment (arrow) called a *vector*. The *direction* of the vector is that of the given quantity and the *length* of the vector is proportional to the magnitude of the quantity.

Example 4.1 An airplane is traveling N40°E at 200 mi/h. Its velocity is represented by the vector **AB** in Figure 4-3.

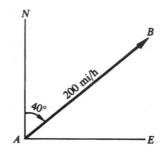

Figure 4-3

Example 4.2 A motor boat having the speed 12 mi/h in still water is headed directly across a river whose current is 4 mi/h. In Figure 4-4, the vector **CD** represents the velocity of the current and the vector **AB** represents, to the same scale, the velocity of the boat in still water. Thus, vector **AB** is three times as long as vector **CD**.

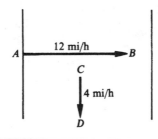

Figure 4-4

Example 4.3 In Figure 4-5, vector **AB** represents a force of 20 lb making an angle of 35° with the positive direction on the x axis and vector **CD** represents a force of 30 lb at 150° with the positive direction on the x axis. Both vectors are drawn to the same scale.

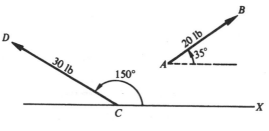

Figure 4-5

Two vectors are said to be equal if they have the same magnitude and direction.

> # Remember
> A vector has no fixed position in a plane and may be moved about in the plane provided only that its magnitude and direction are not changed.

Vector Addition

The *resultant* or *vector sum* of a number of vectors, all in the same plane, is that a vector in the plane would produce the same effect as that produced by all the original vectors acting together.

If two vectors α and β have the same direction, their resultant is a vector **R** whose magnitude is equal to the sum of the magnitudes of the two vectors and whose direction is that of the two vectors. See Figure 4-6(*a*).

If two vectors have opposite directions, their resultant is a vector **R** whose magnitude is the difference (greater magnitude—smaller magnitude) of the magnitudes of the two vectors and whose direction is that of the vector of greater magnitude [Figure 4-6(*b*)].

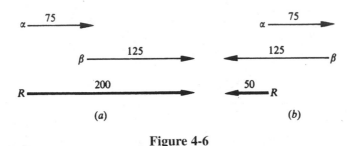

Figure 4-6

In all other cases, the magnitude and direction of the resultant of two vectors is obtained by either of the following two methods:

(1) Parallelogram Method: Place the tail ends of both vectors at any point O in their plane and complete the parallelogram having these vectors as adjacent sides. The directed diagonal issuing from O is the resultant or vector sum of the two given vectors. Thus, in Figure 4-7(*b*), the vector **R** is the resultant of the vectors α and β of Figure 4-7(*a*).

(2) Triangle Method: Choose one of the vectors and label its tail end O. Place the tail end of the other vector at the arrow end of the first. The resultant is then the line segment closing the triangle and directed from O. Thus, in Figures 4-7(*c*) and 4-7(*d*), **R** is the resultant of the vectors α and β.

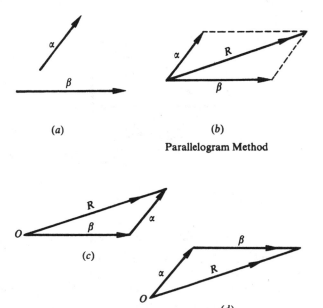

(a)

(b)

Parallelogram Method

(c)

(d)

Triangle Method

Figure 4-7

Components of a Vector

The component of a vector α along a line L is the perpendicular projection of the vector α on L. It is often very useful to resolve a vector into two components along a pair of perpendicular lines.

Example 4.4 In Figure 4-8, the force $\mathbf{F}$ has horizontal component $\mathbf{F}_h =$ $\mathbf{F}\cos 30°$ and vertical component $\mathbf{F}_v = \mathbf{F}\sin 30°$. Note that $\mathbf{F}$ is the vector sum or resultant of $\mathbf{F}_h$ and $\mathbf{F}_v$.

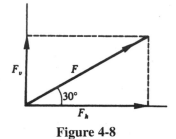

Figure 4-8

Air Navigation

The *heading* of an airplane is the direction (determined from a compass reading) in which the airplane is pointed. The heading is measured clockwise from the north and expressed in degrees and minutes.

The *airspeed* (determined from a reading of the airspeed indicator) is the speed of the airplane in still air.

The *course* (or *track*) of an airplane is the direction in which it moves relative to the ground. The course is measured clockwise from the north.

The *groundspeed* is the speed of the airplane relative to the ground.

The *drift angle* (or wind-correction angle) is the difference (positive) between the heading and the course.

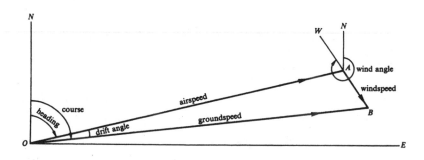

Figure 4-9

In Figure 4-9: ON is the true north line through O
$\angle NOA$ is the heading
OA = the airspeed
AN is the true north line through A
$\angle NAW$ is the wind angle, measured clockwise from
the north line
AB = the windspeed
$\angle NOB$ is the course
OB= the groundspeed
$\angle AOB$ is the drift angle

Note that there are three vectors involved: **OA** representing the airspeed and heading, **AB** representing the direction and speed of the wind, and **OB** representing the groundspeed and course. The groundspeed vector is the resultant of the airspeed vector and the wind vector.

Inclined Plane

An object with weight W on an inclined plane which has an angle of inclination α exerts a force $\mathbf{F}_a$ against the inclined plane and a force $\mathbf{F}_d$ down the inclined plane. The forces $\mathbf{F}_a$ and $\mathbf{F}_d$ are the component vectors for the weight W. See Figure 4-10.

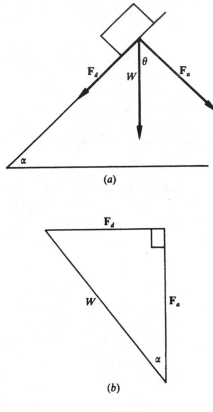

Figure 4-10

Example 4.5 A 500-lb barrel rests on an 11.2° inclined plane. What is the minimum force (ignoring friction) needed to keep the barrel from rolling down the incline and what is the force the barrel exerts against the surface of the inclined plane? (See Figure 4-11.)

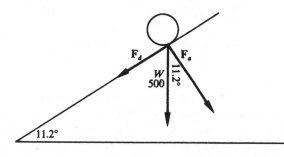

Figure 4-11

$$F_d = 500 \sin 11.2° = 500(0.1942) = 97.1 \text{ lb}$$

$$F_a = 500 \cos 11.2° = 500(0.9810) = 491 \text{ lb}$$

The minimum force needed to keep the barrel from rolling down the incline is 97.1 lb and the force against the inclined plane is 491 lb.

Solved Problem 4.1 A telegraph pole is kept vertical by a guy wire which makes an angle of 25° with the pole and which exerts a pull of F = 300 lb on the top. Find the horizontal and vertical components F_h and F_v of the pull F. See Figure 4-12.

Solution.

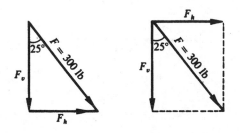

Figure 4-12

$$F_h = 300 \sin 25° = 300(0.4226) = 127 \text{ lb}$$

$$F_v = 300 \cos 25° = 300(0.9063) = 272 \text{ lb}$$

Solved Problem 4.2 A man pulls a rope attached to a sled with a force of 100 lb. The rope makes an angle of 27° with the ground. (*a*) Find the effective pull tending to move the sled along the ground and the effective pull tending to lift the sled vertically. (*b*) Find the force which the man must exert in order that the effective force tending to move the sled along the ground is 100 lb.

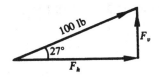

Figure 4-13

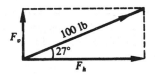

Figure 4-14

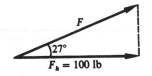

Figure 4-15

Solution.

(*a*) In Figures 4-13 and 4-14, the 100-lb pull in the rope is resolved into horizontal and vertical components, F_h and F_v, respectively. Then

F_h is the force tending to move the sled along the ground and F_v is the force tending to lift the sled.

$F_h = 100 \cos 27° = 100(0.8910) = 89$ lb

$F_v = 100 \sin 27° = 100(0.4540) = 45$ lb

(b) In Figure 4-15, on page 52, the horizontal component of the required force F is $F_h = 100$ lb. Then

$F = 100/\cos 27° = 100/0.8910 = 112$ lb

Solved Problem 4.3 A block weighing $W = 500$ lb rests on a ramp inclined 29° with the horizontal. (a) Find the force tending to move the block down the ramp and the force of the block on the ramp. (b) What minimum force must be applied to keep the block from sliding down the ramp? Neglect friction.

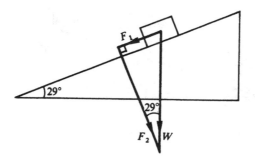

Figure 4-16

Solution.

(a) Refer to Figure 4-16. Resolve the weight W of the block into components F_1 and F_2, respectively, parallel and perpendicular to the ramp. F_1 is the force tending to move the block down the ramp and F_2 is the force of the block on the ramp.

$$F_1 = W \sin 29° = 500(0.4848) = 242 \text{ lb}$$

$$F_2 = W \cos 29° = 500(0.8746) = 437 \text{ lb}$$

(b) 242 lb up the ramp

Solved Problem 4.4 The heading of an airplane is 75° and the airspeed is 200 mi/h. Find the groundspeed and course if there is a wind of 40 mi/h from 165°. Refer to Figure 4-17.

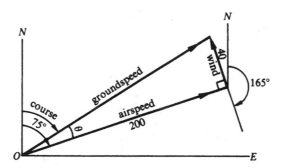

Figure 4-17

Solution.

Construction: Put in the airspeed vector from O, follow through with the wind vector, and close the triangle.

Solution: Groundspeed $= \sqrt{(200)^2 + (40)^2} = 204$ mi/h, $\tan \theta = 40/200 = 0.2000$ and $\theta = 11°20'$, and course $= 75° - \theta = 63°40'$.

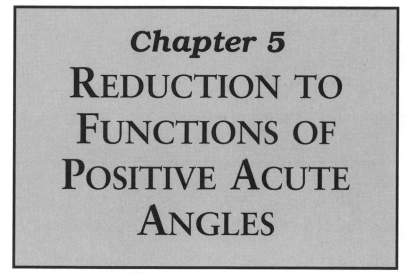

Chapter 5

REDUCTION TO FUNCTIONS OF POSITIVE ACUTE ANGLES

IN THIS CHAPTER:

✔ *Coterminal Angles*
✔ *Functions of a Negative Angle*
✔ *Reference Angles*
✔ *Angles with a Given Function Value*

Coterminal Angles

Let θ be any angle; then

$\sin(\theta + n360°) = \sin\theta$ $\cos(\theta + n360°) = \cos\theta$

$\tan(\theta + n360°) = \tan\theta$ $\cot(\theta + n360°) = \cot\theta$

$\sec(\theta + n360°) = \sec\theta$ $\csc(\theta + n360°) = \csc\theta$

where n is any positive or negative integer or zero.

55

Example 5.1

(*a*) $\sin 400° = \sin (40° + 360°) = \sin 40°$

(*b*) $\cos 850° = \cos (130° + 2 \cdot 360°) = \cos 130°$

(*c*) $\tan (-1000°) = \tan (80° - 3 \cdot 360°) = \tan 80°$

If x is an angle in radian measure, then

$\sin (x + 2n\pi) = \sin x$ $\qquad\qquad$ $\cos (x + 2n\pi) = \cos x$

$\tan (x + 2n\pi) = \tan x$ $\qquad\qquad$ $\cot (x + 2n\pi) = \cot x$

$\sec (x + 2n\pi) = \sec x$ $\qquad\qquad$ $\csc (x + 2n\pi) = \csc x$

where n is any integer.

Functions of a Negative Angle

Let θ be any angle; then

$\sin (-\theta) = -\sin \theta$ $\qquad\qquad$ $\cos (-\theta) = \cos \theta$

$\tan (-\theta) = -\tan \theta$ $\qquad\qquad$ $\cot (-\theta) = -\cot \theta$

$\sec (-\theta) = \sec \theta$ $\qquad\qquad$ $\csc (-\theta) = -\csc \theta$

Reference Angles

If θ is a quadrantal angle, then a reference angle is not needed. Since any angle A can be written as $\theta + n360°$, where n is an integer and $0° \leq \theta \leq 360°$, reference angles will be found for angles from $0°$ to $360°$.

A reference angle R for an angle θ in standard position is the positive acute angle between the x axis and the terminal side of angle θ. The values of the six trigonometric functions of the reference angle for θ, R, agree with the function values for θ except possibly in sign. When the signs of the functions of R are determined by the quadrant of angle θ, then any function of θ can be expressed as a function of the acute angle R. Thus, our tables can be used to find the value of a trigonometric function of any angle.

Quadrant for θ	Relationship	Function Signs
I	$R = \theta$	All functions are positive.
II	$R = 180° - \theta$	Only sin R and csc R positive.
III	$R = \theta - 180°$	Only tan R and cot R positive.
IV	$R = 360° - \theta$	Only cos R and sec R positive.

Angles with a Given Function Value

Since coterminal angles have the same value for a function, there are an unlimited number of angles that have the same value for a trigonometric function. Even when we restrict the angles to the interval of $0°$ to $360°$, there are usually two angles that have the same function value. All the angles that have the same function value also have the same reference angle. The quadrants for the angle are determined by the sign of the function value. The relationships from the previous section are used to find the angle θ, once the reference angle is found.

Solved Problem 5.1 Derive formulas for the functions of $-\theta$ in terms of θ.

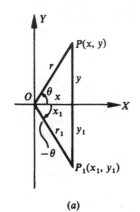

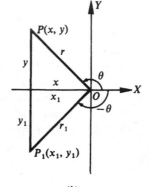

(a)

(b)

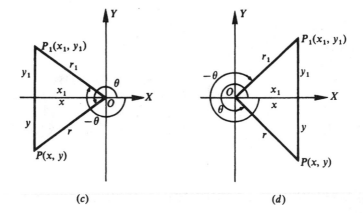

(c)

(d)

Figure 5-1

In Figure 5-1, θ and $-\theta$ are constructed in standard position and are numerically equal. On their respective terminal sides, the points $P(x, y)$ and $P_1(x_1, y_1)$ are located so that $OP = OP_1$. In each of the figures, the two triangles are congruent and $r_1 = r$, $x_1 = x$, and $y_1 = -y$. Then,

$$\sin(-\theta) = \frac{y_1}{r_1} = \frac{-y}{r} = -\frac{y}{r} = -\sin\theta$$

$$\cos(-\theta) = \frac{x_1}{r_1} = \frac{x}{r} = \cos\theta$$

$$\tan(-\theta) = \frac{y_1}{x_1} = \frac{-y}{x} = -\frac{y}{x} = -\tan\theta$$

$$\cot(-\theta) = \frac{x_1}{y_1} = \frac{x}{-y} = -\frac{x}{y} = -\cot\theta$$

$$\sec(-\theta) = \frac{r_1}{x_1} = \frac{r}{x} = \sec\theta$$

$$\csc(-\theta) = \frac{r_1}{y_1} = \frac{r}{-y} = -\frac{r}{y} = -\csc\theta$$

Except for those cases in which a function is not defined, the above relations are also valid when θ is a quadrantal angle. This may be verified by making use of the fact that $-0°$ and $0°$, $-90°$ and $270°$, $-180°$ and $180°$, and $-270°$ and $90°$ are coterminal.

For example, $\sin(-0°) = \sin 0° = 0 = -\sin 0°$, $\sin(-90°) = \sin 270° = -1 = -\sin 90°$, $\cos(-180°) = \cos 180°$, and $\cot(-270°) = \cot 90° = 0 = -\cot 270°$.

Solved Problem 5.2 Verify the equality of the trigonometric functions for θ and its reference angle R where $x > 0$, $y > 0$, and $r = \sqrt{x^2 + y^2}$.

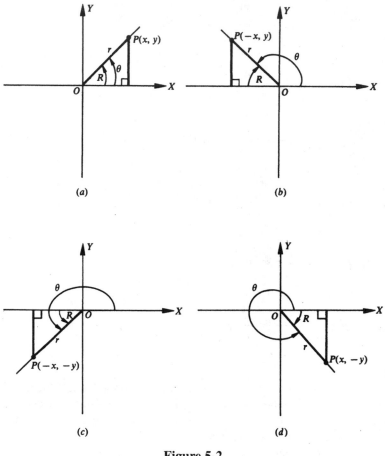

(a)

(b)

(c)

(d)

Figure 5-2

Solution.

(a) θ is in quadrant I. See Figure 5-2(a).

$$\sin \theta = \frac{y}{r} = \sin R \qquad\qquad \cos \theta = \frac{x}{r} = \cos R$$

$$\tan \theta = \frac{y}{x} = \tan R \qquad\qquad \cot \theta = \frac{x}{y} = \cot R$$

$$\sec \theta = \frac{r}{x} = \sec R \qquad\qquad \csc \theta = \frac{r}{y} = \csc R$$

(b) θ is in quadrant II. See Figure 5-2(b).

$$\sin \theta = \frac{y}{r} = \sin R \qquad\qquad \cos \theta = \frac{-x}{r} = -\left(\frac{x}{r}\right) = -\cos R$$

$$\tan \theta = \frac{y}{-x} = -\left(\frac{y}{x}\right) = -\tan R \qquad \cot \theta = \frac{-x}{y} = -\left(\frac{x}{y}\right) = -\cot R$$

$$\sec \theta = \frac{r}{-x} = -\left(\frac{r}{x}\right) = -\sec R \qquad \csc \theta = \frac{r}{y} = \csc R$$

(c) θ is in quadrant III. See Figure 5-2(c).

$$\sin \theta = \frac{-y}{r} = -\left(\frac{y}{r}\right) = -\sin R \qquad \cos \theta = \frac{-x}{r} = -\left(\frac{x}{r}\right) = -\cos R$$

$$\tan \theta = \frac{-y}{-x} = \frac{y}{x} = \tan R \qquad\qquad \cot \theta = \frac{-x}{-y} = \frac{x}{y} = \cot R$$

$$\sec \theta = \frac{r}{-x} = -\left(\frac{r}{x}\right) = -\sec R \qquad \csc \theta = \frac{r}{-y} = -\left(\frac{r}{y}\right) = -\csc R$$

(d) θ is in quadrant IV. See Figure 5-2(d).

$$\sin \theta = \frac{-y}{r} = -\left(\frac{y}{r}\right) = -\sin R \qquad \cos \theta = \frac{x}{r} = \cos R$$

$$\tan \theta = \frac{-y}{x} = -\left(\frac{y}{x}\right) = -\tan R \qquad \cot \theta = \frac{x}{-y} = -\left(\frac{x}{y}\right) = -\cot R$$

$$\sec \theta = \frac{r}{x} = \sec R \qquad \csc \theta = \frac{r}{-y} = -\left(\frac{r}{y}\right) = -\csc R$$

Chapter 6
VARIATIONS AND GRAPHS OF THE TRIGONOMETRIC FUNCTIONS

IN THIS CHAPTER:

✔ *Line Representations of Trigonometric Functions*
✔ *Variations of Trigonometric Functions*
✔ *Graphs of Trigonometric Functions*
✔ *Horizontal and Vertical Shifts*
✔ *Periodic Functions*
✔ *Sine Curves*

Line Representations of Trigonometric Functions

Let θ be any given angle in standard position. (See Figure 6-1 for θ in each of the quadrants.) With the vertex O as center, describe a circle of radius one unit cutting the initial side OX of θ at A, the positive y axis

63

at B, and the terminal side of θ at P. Draw MP perpendicular to OX; draw also the tangents to the circle at A and B meeting the terminal side of θ or its extension through O in the points Q and R, respectively.

In each of the parts of Figure 6-1, the right triangles OMP, OAQ, and OBR are similar, and

$$\sin \theta = \frac{MP}{OP} = MP \qquad\qquad \cos \theta = \frac{OM}{OP} = OM$$

$$\tan \theta = \frac{MP}{OM} = \frac{AQ}{OA} = AQ \qquad\qquad \cot \theta = \frac{OM}{MP} = \frac{BR}{OB} = BR$$

$$\sec \theta = \frac{OP}{OM} = \frac{OQ}{OA} = OQ \qquad\qquad \csc \theta = \frac{OP}{MP} = \frac{OR}{OB} = OR$$

The segments MP, OM, AQ, etc., are directed line segments. The magnitude of a function is given by the length of the corresponding segment and the sign is given by the indicated direction. The directed segments OQ and OR are to be considered positive when measured on the terminal side of the angle and negative when measured on the terminal side extended.

As θ Increases from	0° to 90°	90° to 180°	180° to 270°	270° to 360°
sin θ	*I.* from 0 to 1	*D.* from 1 to 0	*D.* from 0 to −1	*I.* from −1 to 0
cos θ	*D.* from 1 to 0	*D.* from 0 to −1	*I.* from −1 to 0	*I.* from 0 to 1
tan θ	*I.* from 0 without limit (0 to +∞)	*I.* from large negative values to 0 (−∞ to 0)	*I.* from 0 without limit (0 to +∞)	*I.* from large negative values to 0 (−∞ to 0)
cot θ	*D.* from large positive values to 0 (+∞ to 0)	*D.* from 0 without limit (0 to −∞)	*D.* from large positive values to 0 (+∞ to 0)	*D.* from 0 without limit (0 to −∞)
sec θ	*I.* from 1 without limit (1 to +∞)	*I.* from large negative values to −1 (−∞ to −1)	*D.* from −1 without limit (−1 to −∞)	*D.* from large positive values to 1 (+∞ to 1)
csc θ	*D.* from large positive values to 1 (+∞ to 1)	*I.* from 1 without limit (1 to +∞)	*I.* from large negative values to −1 (−∞ to −1)	*D.* from −1 without limit (−1 to −∞)

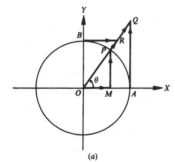

(a)

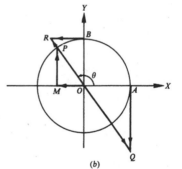

(b)

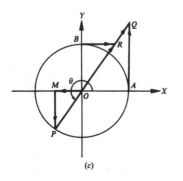

(c)

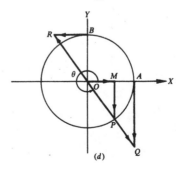

(d)

Figure 6-1

Variations of Trigonometric Functions

Let P move counterclockwise about the unit circle, starting at A, so that $\theta = \angle AOP$ varies continuously from $0°$ to $360°$. Using Figure 6-1, see how the trigonometric functions vary (I = increases, D = decreases).

Graphs of Trigonometric Functions

In the following table, values of the angle x are given in radians. Whenever a trigonometric function is undefined for the value of x, $\pm\infty$ is recorded instead of a function value. The graphs of the trigonometric functions are shown in Figure 6-2.

x	$y = \sin x$	$y = \cos x$	$y = \tan x$	$y = \cot x$	$y = \sec x$	$y = \csc x$
0	0	1.00	0	$\pm\infty$	1.00	$\pm\infty$
$\pi/6$	0.50	0.87	0.58	1.73	1.15	2.00
$\pi/4$	0.71	0.71	1.00	1.00	1.41	1.41
$\pi/3$	0.87	0.50	1.73	0.58	2.00	1.15
$\pi/2$	1.00	0	$\pm\infty$	0	$\pm\infty$	1.00
$2\pi/3$	0.87	-0.50	-1.73	-0.58	-2.00	1.15
$3\pi/4$	0.71	-0.71	-1.00	-1.00	-1.41	1.41
$5\pi/6$	0.50	-0.87	-0.58	-1.73	-1.15	2.00
π	0	-1.00	0	$\pm\infty$	-1.00	$\pm\infty$
$7\pi/6$	-0.50	-0.87	0.58	1.73	-1.15	-2.00
$5\pi/4$	-0.71	-0.71	1.00	1.00	-1.41	-1.41
$4\pi/3$	-0.87	-0.50	1.73	0.58	-2.00	-1.15
$3\pi/2$	-1.00	0	$\pm\infty$	0	$\pm\infty$	-1.00
$5\pi/3$	-0.87	0.50	-1.73	-0.58	2.00	-1.15
$7\pi/4$	-0.71	0.71	-1.00	-1.00	1.41	-1.41
$11\pi/6$	-0.50	0.87	-0.58	-1.73	1.15	-2.00
2π	0	1.00	0	$\pm\infty$	1.00	$\pm\infty$

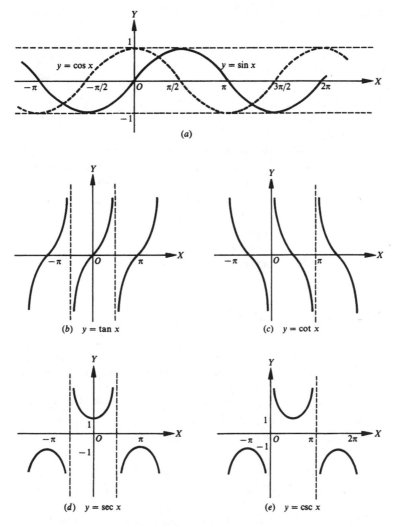

(a)

(b) $y = \tan x$

(c) $y = \cot x$

(d) $y = \sec x$

(e) $y = \csc x$

Figure 6-2

Horizontal and Vertical Shifts

The graph of a trigonometric function can be shifted vertically by adding a nonzero constant to the function and horizontally by adding a nonzero constant to the angle of the trigonometric function. Figure 6-3(*a*) is the graph of $y = \sin x$ and the remaining parts of Figure 6-3 are the results of shifting this graph.

If c is a positive number, then adding it to a trigonometric function results in the graph being shifted up c units [see Figure 6-3(*b*)] and subtracting it from a trigonometric function results in the graph being shifted down c units [see Figure 6-3(*c*)].

For a positive number d, a trigonometric function is shifted left d units when d is added to the angle [see Figure 6-3(*d*)] and shifted right d units when d is subtracted from the angle [see Figure 6-3(*e*)].

Periodic Functions

Any function of a variable x, $f(x)$, which repeats its values in definite cycles is called *periodic*. The smallest range of values of x, which corresponds to a complete cycle of values of the function, is called the *period* of the function. It is evident from the graphs of the trigonometric functions that the sine, cosine, secant, and cosecant are of period 2π while the tangent and cotangent are of period π.

1. Since $\sin (\tfrac{1}{2}\pi + x) = \cos x$, the graph of $y = \cos x$ may be obtained most easily by shifting the graph of $y = \sin x$ a distance $\tfrac{1}{2}\pi$ to the left.
2. Since $\csc (\tfrac{1}{2}\pi + x) = \sec x$, the graph of $y = \csc x$ may be obtained by shifting the graph of $y = \sec x$ a distance $\tfrac{1}{2}\pi$ to the right.

(a) $y = \sin x$

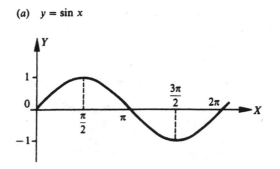

(b) $y = \sin x + 2$

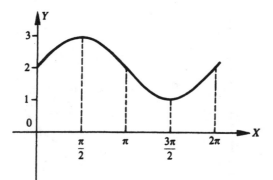

(c) $y = \sin x - 1$

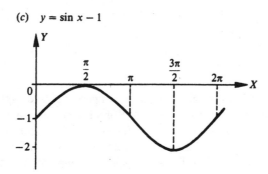

Figure 6-3

(d) $y = \sin(x + \pi/4)$

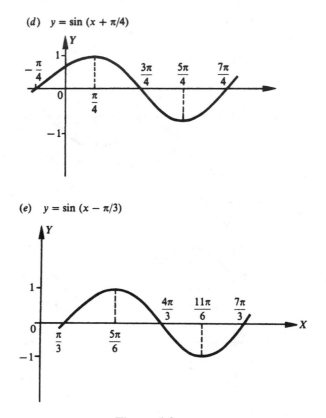

(e) $y = \sin(x - \pi/3)$

Figure 6-3

Sine Curves

The *amplitude* (maximum ordinate) and period (wavelength) of $y = \sin x$ are respectively 1 and 2π. For a given value of x, the value of $y = a \sin x$, $a > 0$, is a times the value of $y = \sin x$. Thus, the amplitude of $y = a \sin x$ is a and the period is 2π. Since when $bx = 2\pi$, $x = 2\pi/b$, the amplitude of $y = \sin bx$, $b > 0$, is 1 and the period is $2\pi/b$.

The general sine curve (sinusoid) of equation

$$y = a \sin bx, \quad a > 0, b > 0$$

has amplitude a and period $2\pi/b$. Thus, the graph of $y = 3 \sin 2x$ has amplitude 3 and period $2\pi/2 = \pi$. Figure 6-4 exhibits the graphs of $y = \sin x$ and $y = 3 \sin 2x$ on the same axis.

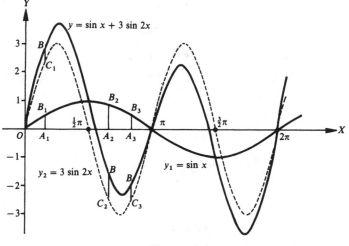

Figure 6-4

More complicated forms of wave motions are obtained by combining two or more sine curves.

Solved Problem 6.1 Sketch the graphs of the following for one period: (a) $y = 4 \sin x$; (b) $y = \sin 3x$; (c) $y = 3 \sin (\frac{1}{2})x$; (d) $y = 2 \cos x = 2 \sin (x + (\frac{1}{2})\pi)$; (e) $y = 3 \cos (\frac{1}{2})x = 3 \sin ((\frac{1}{2})x + (\frac{1}{2})\pi)$.

Solution. In each case, we use the same curve and then put in the y axis and choose the units on each axis to satisfy the requirements of amplitude and period of each curve (see Figure 6-5).

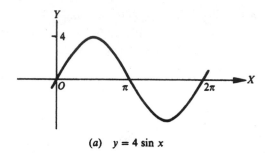

(a) $y = 4 \sin x$

(b) $y = \sin 3x$

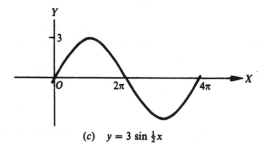

(c) $y = 3 \sin \frac{1}{2}x$

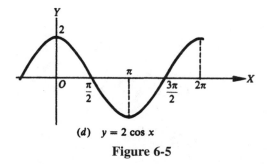

(d) $y = 2 \cos x$

Figure 6-5

(*a*) $y = 4 \sin x$ has amplitude = 4 and period = 2π.

(*b*) $y = \sin 3x$ has amplitude = 1 and period = $2\pi/3$

(*c*) $y = 3 \sin (½)x$ has amplitude = 3 and period = $2\pi/(½) = 4\pi$.

(*d*) $y = 2 \cos x$ has amplitude = 2 and period = 2π. Note the position of the *y* axis.

(*e*) $y = 3 \cos (½)x$ has amplitude = 3 and period = 4π.

Solved Problem 6.2 Construct the graph of each of the following: (*a*) $y = ½ \tan x$; (*b*) $y = 3 \tan x$; (*c*) $y = \tan 3x$; (*d*) $y = \tan (¼)x$.

Solution. In each case, we use the same curve and then put in the *y* axis and choose the units on the *x* axis to satisfy the period of the curve (see Figure 6-6).

(*a*) $y = \tfrac{1}{2} \tan x$ has period π (*b*) $y = 3 \tan x$ has period π

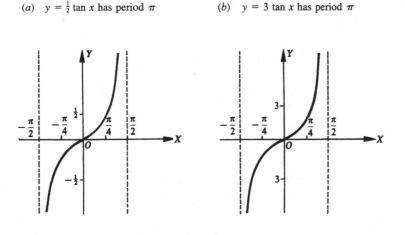

Figure 6-6

(c) $y = \tan 3x$ has period $\pi/3$ (d) $y = \tan \frac{1}{4}x$ has period $\pi/\frac{1}{4} = 4\pi$

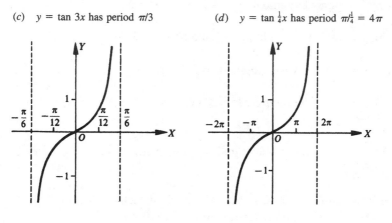

Figure 6-6, cont.

Solved Problem 6.3 Construct the graph of each of the following: (a) $y = \sin x + \cos x$; (b) $y = \sin 2x + \cos 3x$; (c) $y = \sin 2x - \cos 3x$; (d) $y = 3 \sin 2x + 2 \cos 3x$.

Solution.

(a) $y = \sin x + \cos x$

Figure 6-7(a)

(b) $y = \sin 2x + \cos 3x$

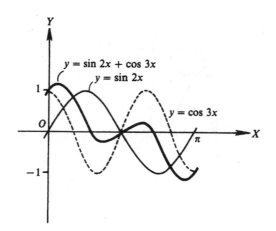

Figure 6-7(b)

(c) $y = \sin 2x - \cos 3x$

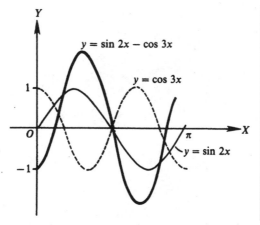

Figure 6-7(c)

(d) $y = 3 \sin 2x + 2 \cos 3x$

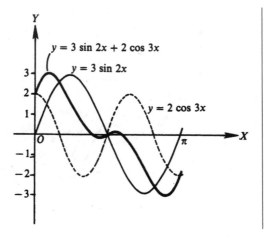

Figure 6-7(d)

Chapter 7
BASIC RELATIONSHIPS AND IDENTITIES

In This Chapter:

✔ Basic Relationships
✔ Simplification of Trigonometric Expressions
✔ Trigonometric Identities

Basic Relationships

Reciprocal Relationships	Quotient Relationships	Pythagorean Relationships
$\csc \theta = \dfrac{1}{\sin \theta}$	$\tan \theta = \dfrac{\sin \theta}{\cos \theta}$	$\sin^2 \theta + \cos^2 \theta = 1$
$\sec \theta = \dfrac{1}{\cos \theta}$	$\cot \theta = \dfrac{\cos \theta}{\sin \theta}$	$1 + \tan^2 \theta = \sec^2 \theta$

$$\cot \theta = \frac{1}{\tan \theta}$$
$$1 + \cot^2 \theta = \csc^2 \theta$$

The basic relationships hold for every value of θ for which the functions involved are defined.

Thus, $\sin^2 \theta + \cos^2 \theta = 1$ holds for every value of θ, while $\tan \theta = \sin \theta / \cos \theta$ holds for all values of θ for which $\tan \theta$ is defined, i.e., for all $\theta \neq n \cdot 90°$ where n is odd. Note that for the excluded values of θ, $\cos \theta = 0$ and $\sin \theta \neq 0$.

Simplification of Trigonometric Expressions

It is frequently desirable to transform or reduce a given expression involving trigonometric functions to a simpler form.

Solved Problem 7.1   Using the relation $\sin^2 \theta + \cos^2 \theta = 1$, simplify the trigonometric expression $\sin^3 \theta + \sin \theta \cos^2 \theta$.

Solution.

$$\sin^3\theta + \sin \theta \cos^2 \theta = \sin \theta (\sin^2 \theta + \cos^2 \theta) = \sin \theta (1) = \sin \theta$$

Trigonometric Identities

An equation involving the trigonometric functions that is valid for all values of the angle for which the functions are defined is called a *trigonometric identity*. Examples are:

$$\cos \theta \csc \theta = \cot \theta \qquad \text{and} \qquad \cos \theta \tan \theta = \sin \theta$$

A trigonometric identity is verified by transforming one member (your choice) into the other. In general, one begins with the more complicated side. In some cases, each side is transformed into the same new form.

General Guidelines for Verifying Identities

1. Know the eight basic relationships and recognize alternative forms of each.

2. Know the procedures for adding and subtracting fractions, reducing fractions, and transforming fractions into equivalent fractions.

3. Know factoring and special product techniques.

4. Use only substitution and simplification procedures that allow you to work on exactly one side of an equation.

5. Select the side of the equation that appears more complicated and attempt to transform it into the form of the other side of the equation.

6. Transform each side of the equation, independently, into the same form.

7. Avoid substitutions that introduce radicals.

8. Use substitutions to change all trigonometric functions into expressions involving only sine and cosine and then simplify.

9. Multiply the numerator and denominator of a fraction by the conjugate of either.

10. Simplify a square root of a fraction by using conjugates to transform it into the quotient of perfect squares.

Solved Problem 7.2 Prove the Pythagorean relationships (a) $\sin^2 \theta + \cos^2 \theta = 1$, (b) $1 + \tan^2 \theta = \sec^2 \theta$, and (c) $1 + \cot^2 \theta = \csc^2 \theta$.

Solution. For $P(x, y)$, we have $A = (x^2 + y^2 = r^2)$.

(a) Dividing A by r^2, $(x/r)^2 + (y/r)^2 = 1$ and $\sin^2 \theta + \cos^2 \theta = 1$.

(b) Dividing A by x^2, $1 + (y/x)^2 = (r/x)^2$ and $1 + \tan^2 \theta = \sec^2 \theta$.

Also, dividing $\sin^2 \theta + \cos^2 \theta = 1$ by $\cos^2 \theta$,

$$\left(\frac{\sin \theta}{\cos \theta} \right)^2 + 1 = \left(\frac{1}{\cos \theta} \right)^2 \quad \text{or} \quad \tan^2 \theta + 1 = \sec^2 \theta.$$

(c) Dividing A by y^2, $+ (x/y)^2 + 1 = (r/y)^2$ and $\cot^2 \theta + 1 = \csc^2 \theta$.

Also, dividing $\sin^2 \theta + \cos^2 \theta = 1$ by $\sin^2 \theta$,

$$1 + \left(\frac{\cos \theta}{\sin \theta} \right)^2 = \left(\frac{1}{\sin \theta} \right)^2 \quad \text{or} \quad 1 + \cot^2 \theta = \csc^2 \theta.$$

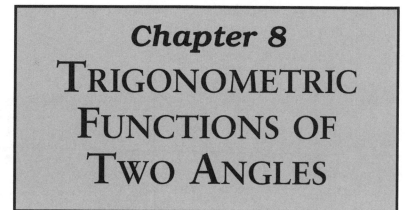

Chapter 8
TRIGONOMETRIC FUNCTIONS OF TWO ANGLES

IN THIS CHAPTER:

✔ *Addition Formulas*
✔ *Subtraction Formulas*
✔ *Double-Angle Formulas*
✔ *Half-Angle Formulas*

Addition Formulas

$$\sin(\alpha + \beta) = \sin \alpha \cos \beta + \cos \alpha \sin \beta$$

$$\cos(\alpha + \beta) = \cos \alpha \cos \beta - \sin \alpha \sin \beta$$

$$\tan(\alpha + \beta) = \frac{\tan \alpha + \tan \beta}{1 - \tan \alpha \ \tan \beta}$$

Solved Problem 8.1 Prove the addition formulas.

Solution.   Let α and β be positive acute angles such that $\alpha + \beta < 90^\circ$ [Figure 8-1(a)] or $\alpha + \beta > 90^\circ$ [Figure 8-1(b)].

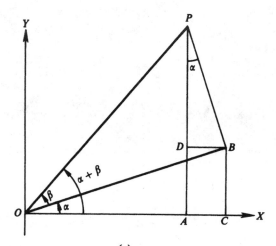

(a)

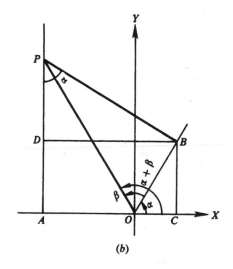

(b)

Figure 8-1

To construct these figures, place angle α in standard position and then place angle β with its vertex at O and its initial side along the terminal side of angle α. Let P be any point on the terminal side of angle (α + β). Draw PA perpendicular to OX, PB perpendicular to the terminal side of angle α, BC perpendicular to OX, and BD perpendicular to AP.

Now $\angle APB = \alpha$ since corresponding sides (OA and AP, and OB and BP) are perpendicular. Then,

$\sin(\alpha + \beta) \ =$

$$\frac{AP}{OP} = \frac{AD + DP}{OP} = \frac{CB + DP}{OP} = \frac{CB}{OP} + \frac{DP}{OP} = \frac{CB}{OB} \cdot \frac{OB}{OP} + \frac{DP}{BP} \cdot \frac{BP}{OP}$$

$$= \sin \alpha \cos \beta \ + \cos \alpha \sin \beta$$

$\cos(\alpha + \beta) \ =$

$$\frac{OA}{OP} = \frac{OC - AC}{OP} = \frac{OC - DB}{OP} = \frac{OC}{OP} - \frac{DB}{OP} = \frac{OC}{OB} \cdot \frac{OB}{OP} - \frac{DB}{BP} \cdot \frac{BP}{OP}$$

$$= \cos \alpha \cos \beta \ - \sin \alpha \sin \beta$$

$\tan(\alpha + \beta) \ =$

$$\frac{\sin(\alpha + \beta)}{\cos(\alpha + \beta)} = \frac{\sin \alpha \cos \beta + \cos \alpha \sin \beta}{\cos \alpha \cos \beta - \sin \alpha \sin \beta} = \frac{\dfrac{\sin \alpha \cos \beta}{\cos \alpha \cos \beta} + \dfrac{\cos \alpha \sin \beta}{\cos \alpha \cos \beta}}{\dfrac{\cos \alpha \cos \beta}{\cos \alpha \cos \beta} - \dfrac{\sin \alpha \sin \beta}{\cos \alpha \cos \beta}}$$

$$= \frac{\tan \alpha + \tan \beta}{1 - \tan \alpha \ \tan \beta}$$

Subtraction Formulas

$$\sin (\alpha - \beta) = \sin \alpha \cos \beta - \cos \alpha \sin \beta$$

$$\cos (\alpha - \beta) = \cos \alpha \cos \beta + \sin \alpha \sin \beta$$

$$\tan (\alpha - \beta) = \frac{\tan \alpha - \tan \beta}{1 + \tan \alpha \ \tan \beta}$$

Solved Problem 8.2 Prove the subtraction formulas.

Solution.

$$\sin (\alpha - \beta) \ = \sin [\alpha + (-\beta)] = \sin \alpha \cos (-\beta) + \cos \alpha \sin (-\beta)$$

$$= \sin \alpha (\cos \beta) + \cos \alpha (-\sin \beta)$$

$$= \sin \alpha \cos \beta - \cos \alpha \sin \beta$$

$$\cos (\alpha - \beta) \ = \cos [\alpha + (-\beta)] = \cos \alpha \cos (-\beta) - \sin \alpha \sin (-\beta)$$

$$= \cos \alpha (\cos \beta) - \sin \alpha (-\sin \beta)$$

$$= \cos \alpha \cos \beta + \sin \alpha \sin \beta$$

$$\tan (\alpha - \beta) \ = \tan [\alpha + (-\beta)] = \frac{\tan \alpha + \tan(-\beta)}{1 - \tan \alpha \ \tan(-\beta)}$$

$$= \frac{\tan \alpha + (-\tan \beta)}{1 - \tan \alpha \ (-\tan \beta)}$$

$$= \frac{\tan \alpha - \tan \beta}{1 + \tan \alpha \ \tan \beta}$$

Double-Angle Formulas

$$\sin 2\alpha = 2 \sin \alpha \cos \alpha$$

$$\cos 2\alpha = \cos^2 \alpha - \sin^2 \alpha = 1 - 2 \sin^2 \alpha = 2 \cos^2 \alpha - 1$$

$$\tan 2\alpha = \frac{2 \tan \alpha}{1 - \tan^2 \alpha}$$

Solved Problem 8.3 Prove the double-angle formulas.

Solution.   In each of the addition formulas

$$\sin (\alpha + \beta) = \sin \alpha \cos \beta + \cos \alpha \sin \beta$$

$$\cos (\alpha + \beta) = \cos \alpha \cos \beta - \sin \alpha \sin \beta$$

$$\tan (\alpha + \beta) = \frac{\tan \alpha + \tan \beta}{1 - \tan \alpha \ \tan \beta},$$

replace β with α. Then,

$$\sin 2\alpha = \sin \alpha \cos \alpha + \cos \alpha \sin \alpha = 2 \sin \alpha \cos \alpha$$

$$\cos 2\alpha = \cos \alpha \cos \alpha - \sin \alpha \sin \alpha$$

$$= \cos^2 \alpha - \sin^2 \alpha$$

$$= (1 - \sin^2 \alpha) - \sin^2 \alpha = 1 - 2 \sin^2 \alpha$$

$$= \cos^2 \alpha - (1 - \cos^2 \alpha) = 2 \cos^2 \alpha - 1$$

$$\tan 2\alpha = \frac{\tan \alpha + \tan \alpha}{1 - \tan \alpha \tan \alpha} = \frac{2 \tan \alpha}{1 - \tan^2 \alpha}$$

Half-Angle Formulas

$$\sin \tfrac{1}{2}\theta = \pm \sqrt{\frac{1 - \cos \theta}{2}}$$

$$\cos \tfrac{1}{2}\theta = \pm \sqrt{\frac{1 + \cos \theta}{2}}$$

$$\tan \tfrac{1}{2}\theta = \pm \sqrt{\frac{1 - \cos \theta}{1 + \cos \theta}} = \frac{\sin \theta}{1 + \cos \theta} = \frac{1 - \cos \theta}{\sin \theta}$$

Solved Problem 8.4   Prove the half-angle formulas.

Solution.

In $\cos 2\alpha = 1 - 2 \sin^2 \alpha$, replace α with $\tfrac{1}{2}\theta$. Then,

$$\cos 2\alpha = 1 - 2 \sin^2 \tfrac{1}{2}\theta$$

$$\sin^2 \tfrac{1}{2}\theta = \frac{1 - \cos \theta}{2}$$

$$\sin \tfrac{1}{2}\theta = \pm \sqrt{\frac{1 - \cos \theta}{2}}$$

In $\cos 2\alpha = 2\cos^2 \alpha - 1$, replace α with $\frac{1}{2}\theta$. Then,

$$\cos 2\alpha = 2\cos^2 \tfrac{1}{2}\theta - 1$$

$$\cos^2 \tfrac{1}{2}\theta = \frac{1+\cos\theta}{2}$$

$$\cos \tfrac{1}{4}\theta = \pm\sqrt{\frac{1+\cos\theta}{2}}$$

Finally,

$$\tan \tfrac{1}{2}\theta = \frac{\sin\dfrac{1}{2}\theta}{\cos\dfrac{1}{2}\theta} = \pm\sqrt{\frac{1-\cos\theta}{1+\cos\theta}}$$

$$= \pm\sqrt{\frac{(1-\cos\theta)(1+\cos\theta)}{(1+\cos\theta)(1+\cos\theta)}} = \pm\sqrt{\frac{1-\cos^2\theta}{(1+\cos\theta)^2}} = \frac{\sin\theta}{1+\cos\theta}$$

$$= \pm\sqrt{\frac{(1-\cos\theta)(1-\cos\theta)}{(1+\cos\theta)(1-\cos\theta)}} = \pm\sqrt{\frac{(1-\cos\theta)^2}{1-\cos^2\theta}} = \frac{1-\cos\theta}{\sin\theta}$$

The signs $\pm$ are not needed here since $\tan \frac{1}{4}\theta$ and $\sin\theta$ always have the same sign and $1 - \cos\theta$ is always positive.

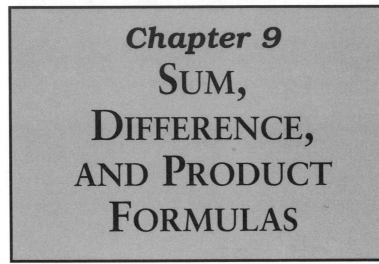

Chapter 9
SUM, DIFFERENCE, AND PRODUCT FORMULAS

IN THIS CHAPTER:

 Products of Sines and Cosines
 Sum and Difference of Sines and Cosines

Products of Sines and Cosines

$$\sin \alpha \cos \beta = \tfrac{1}{2}[\sin(\alpha + \beta) + \sin(\alpha - \beta)]$$

$$\cos \alpha \sin \beta = \tfrac{1}{2}[\sin(\alpha + \beta) - \sin(\alpha - \beta)]$$

$$\cos \alpha \cos \beta = \tfrac{1}{2}[\cos(\alpha + \beta) + \cos(\alpha - \beta)]$$

$$\sin \alpha \sin \beta = -\tfrac{1}{2}[\cos(\alpha + \beta) - \cos(\alpha - \beta)]$$

Solved Problem 9.1 Derive the product formulas.

Solution.

Since

$$\sin(\alpha + \beta) + \sin(\alpha - \beta)$$

$$= (\sin \alpha \cos \beta + \cos \alpha \cos \beta) + (\sin \alpha \cos \beta - \cos \alpha \sin \beta)$$

$$= 2 \sin\alpha \cos \beta$$

Thus,

$$\sin\alpha \cos \beta = \tfrac{1}{2}[\sin(\alpha + \beta) + \sin(\alpha - \beta)]$$

Since

$$\sin(\alpha + \beta) - \sin(\alpha - \beta) = 2 \cos\alpha \sin \beta$$

Thus,

$$\cos\alpha \sin \beta = \tfrac{1}{2}[\sin(\alpha + \beta) - \sin(\alpha - \beta)]$$

Since

$$\cos(\alpha + \beta) + \cos(\alpha - \beta)$$

$$= (\cos\alpha \cos \beta - \sin\alpha \sin \beta) + (\cos\alpha \cos \beta - \sin\alpha \sin \beta)$$

$$= 2 \cos \alpha \cos \beta$$

Thus,

$$\cos\alpha \cos \beta = \tfrac{1}{4}[\cos(\alpha + \beta) + \sin(\alpha - \beta)]$$

Since

$$\cos(\alpha + \beta) - \cos(\alpha - \beta) = -2 \sin\alpha \sin \beta$$

Thus,

$$\sin \alpha \sin \beta = -\tfrac{1}{2}[\cos(\alpha + \beta) - \cos(\alpha - \beta)]$$

Sum and Difference of Sines and Cosines

$$\sin A + \sin B = 2 \sin \tfrac{1}{2}(A + B) \cos \tfrac{1}{2}(A - B)$$

$$\sin A - \sin B = 2 \cos \tfrac{1}{2}(A + B) \sin \tfrac{1}{2}(A - B)$$

$$\cos A + \cos B = 2 \cos \tfrac{1}{2}(A + B) \cos \tfrac{1}{2}(A - B)$$

$$\cos A - \cos B = -2 \sin \tfrac{1}{2}(A + B) \sin \tfrac{1}{2}(A - B)$$

Solved Problem 9.2 Derive the sum and difference formulas.

Solution. Let $\alpha + \beta = A$ and $\alpha - \beta = B$ so that $\alpha = \tfrac{1}{2}(A + B)$ and $\beta = \tfrac{1}{2}(A - B)$. Then,

$$\sin (\alpha + \beta) + \sin (\alpha - \beta) = 2 \sin \alpha \cos \beta$$

becomes $\sin A + \sin B = 2 \sin \tfrac{1}{2}(A + B) \cos \tfrac{1}{2}(A - B)$

$$\sin (\alpha + \beta) - \sin (\alpha - \beta) = 2 \cos \alpha \sin \beta$$

becomes $\sin A - \sin B = 2 \cos \tfrac{1}{2}(A + B) \sin \tfrac{1}{2}(A - B)$

$\cos (\alpha + \beta) + \cos (\alpha - \beta) = 2 \cos \alpha \cos \beta$

  becomes $\cos A + \cos B = 2 \cos \frac{1}{2}(A + B) \cos \frac{1}{2}(A - B)$

$\cos (\alpha + \beta) - \cos (\alpha - \beta) = -2 \sin \alpha \cos \beta$

  becomes $\cos A - \cos B = -2 \sin \frac{1}{2}(A + B) \sin \frac{1}{2}(A - B)$

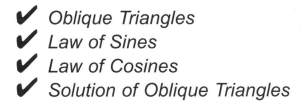

Chapter 10
OBLIQUE TRIANGLES

IN THIS CHAPTER:

✔ *Oblique Triangles*
✔ *Law of Sines*
✔ *Law of Cosines*
✔ *Solution of Oblique Triangles*

Oblique Triangles

An *oblique triangle* is one which does not contain a right angle. Such a triangle contains either three acute angles or two acute angles and one obtuse angle.

The convention of denoting the angles by *A*, *B*, and *C* and the lengths of the corresponding opposite sides by *a*, *b*, and *c* will be used here. See Figure 10-1.

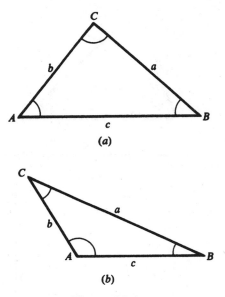

(a)

(b)

Figure 10-1

Law of Sines

In any triangle ABC, the ratio of a side and the sine of the opposite angle is a constant; i.e.,

$$\frac{a}{\sin A} = \frac{b}{\sin B} = \frac{c}{\sin C} \qquad \text{or} \qquad \frac{\sin A}{a} = \frac{\sin B}{b} = \frac{\sin C}{c}$$

Solved Problem 10.1 Derive the law of sines.

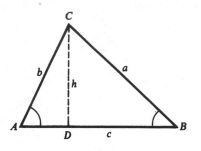

Figure 10-2

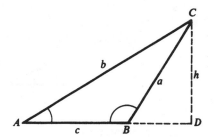

Figure 10-3

Solution.

Let ABC be any oblique triangle. In Figure 10-2, angles A and B are acute, while in Figure 10-3, angle B is obtuse. Draw CD perpendicular to AB or AB extended and denote its length by h.

In the right triangle ACD of either figure, $h = b \sin A$, while in the right triangle BCD, $h = a \sin B$ since in Figure 10-3, $h = a \sin \angle DBC = a \sin (180° - B) = a \sin B$. Thus,

$$a \sin B = b \sin A \qquad \text{or} \qquad \frac{a}{\sin A} = \frac{b}{\sin B}$$

In a similar manner (by drawing a perpendicular from B to AC or a perpendicular from A to BC), we obtain

$$\frac{a}{\sin A} = \frac{c}{\sin C} \qquad \text{or} \qquad \frac{b}{\sin B} = \frac{c}{\sin C}$$

Thus, finally,

$$\frac{a}{\sin A} = \frac{b}{\sin B} = \frac{c}{\sin C}$$

Law of Cosines

In any triangle ABC, the square of any side is equal to the sum of the squares of the other two sides diminished by twice the product of these sides and the cosine of the included angle; i.e.,

$$a^2 = b^2 + c^2 - 2bc \cos A$$

$$b^2 = a^2 + c^2 - 2ac \cos B$$

$$c^2 = a^2 + b^2 - 2ab \cos C$$

Solved Problem 10.2 Derive the law of cosines.

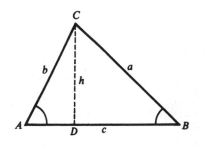

Figure 10-4

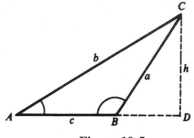

Figure 10-5

Solution.

In either right triangle ACD of Figures 10-4 and 10-5, $b^2 = h^2 + (AD)^2$.
In the right triangle BCD of Figure 10-4, $h = a \sin B$ and $DB = a \cos B$. Then,

$AD = AB - DB = c - a \cos B$

and

$$b^2 = h^2 + (AD)^2 = a^2\sin^2 B + c^2 - 2ca \cos B + a^2\cos^2 B$$

$$= a^2(\sin^2 B + \cos^2 B) + c^2 - 2ca \cos B$$

$$= a^2 + c^2 - 2ca \cos B$$

In the right triangle BCD of Figure 10-5,

$$h = a \sin \angle CBD = a \sin (180° - B) = a \sin B.$$

$$BD = a \cos \angle CBD = a \cos (180° - B) = - a \cos B.$$

Then,

$$AD = AB + BD = c - a \cos B$$

and

$$b^2 = a^2 + c^2 - 2ac \cos B$$

The remaining equations may be obtained by cyclic changes of the letters.

Solution of Oblique Triangles

When three parts of a triangle, not all angles, are known, the triangle is uniquely determined, except in one case noted below. The five cases of oblique triangles are:

- Case I: Given two angles and the side opposite one of them, the law of sines is used to find the side opposite the second given angle.

Solved Problem 10.3 Solve the triangle ABC, given $a = 62.5$, $A = 112°20'$, and $C = 42°10'$. See Figure 10-6.

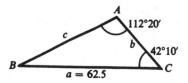

Figure 10-6

Solution.

For B: $B = 180° - (C + A) = 180° - 154°30' = 25°30'$

For b: $b = \dfrac{a \sin B}{\sin A} = \dfrac{62.5 \sin 25°30'}{\sin 112°20'} = \dfrac{62.5(0.4305)}{0.9250} = 29.1$

$$[\sin 112°20' = \sin(180° - 112°20') = \sin 67°40']$$

For c: $c = \dfrac{a \sin C}{\sin A} = \dfrac{62.5 \sin 42°10'}{\sin 112°20'} = \dfrac{62.5(0.6713)}{0.9250} = 45.4$

The required parts are $b = 29.1$, $c = 45.4$, and $B = 25°30'$.

- **Case II: Given two angles and the included side**, the law of sines is used to find the third angle, then either of the remaining sides

Solved Problem 10.4 Solve the triangle ABC, given $c = 25$, $A = 35°$, and $B = 68°$. See Figure 10-7.

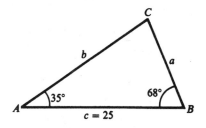

Figure 10-7

Solution.

For C: $C = 180° - (A + B) = 180° - 103° = 77°$

For a: $a = \dfrac{c \sin A}{\sin C} = \dfrac{25 \sin 35°}{\sin 77°} = \dfrac{25(0.5736)}{0.9744} = 15$

For b: $$b = \frac{c \sin B}{\sin C} = \frac{25 \sin 68°}{\sin 77°} = \frac{25(0.9272)}{0.9744} = 24$$

The required parts are $a = 15$, $b = 24$, and $B = 77°$.

- Case III: Given two sides and the angle opposite one of them, the law of sines is used to find the angle opposite the second given side.

Solved Problem 10.5 Solve the triangle ABC, given $a = 525$, $c = 421$, and $A = 130°50'$. See Figure 10-8.

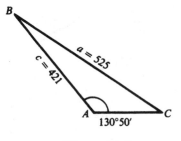

Figure 10-8

Solution. Since A is obtuse and $a > c$, there is one solution.

For C:

$$\sin C = \frac{c \sin A}{a} = \frac{421 \sin 130°50'}{525} = \frac{421(0.7566)}{525} = 0.6067 \text{ and } C = 37°20'$$

For B: $B = 180° - (C + A) = 11°50'$

For b: $b = \dfrac{a \sin B}{\sin A} = \dfrac{525 \sin 11°50'}{\sin 130°50'} = \dfrac{525(0.2051)}{0.7566} = 142$

The required parts are $C = 37°20'$, $B = 11°50'$, and $b = 142$.

- Case IV: Given two sides and the included angle, the law of cosines is used to find the third side.

Solved Problem 10.6 Solve the triangle ABC, given $a = 132$, $b = 224$, and $C = 28°40'$. See Figure 10-9.

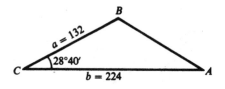

Figure 10-9

Solution.

For c: $c^2 = a^2 + b^2 - 2ab \cos C$

$\quad\quad = (132)^2 + (224)^2 - 2(132)(224) \cos 28°40'$

$\quad\quad = (132)^2 + (224)^2 - 2(132)(224)(0.8774)$

$\quad\quad = 15{,}714$

$\quad c = 125$

For A:

$$\sin A =$$
$$\frac{a \sin C}{c} = \frac{132 \sin 28°40'}{125} = \frac{132(0.4797)}{125} = 0.5066 \text{ and } A = 30°30'$$

For B:

$$\sin B =$$
$$\frac{b \sin C}{c} = \frac{224 \sin 28°40'}{125} = \frac{224(0.4797)}{125} = 0.8596 \text{ and } B = 120°40'$$

(Since $b > a$, A is acute; since $A + C < 90°$, $B > 90°$.)

Check: $A + B + C = 179°50'$.

The required parts are $A = 30°30'$, $B = 120°40'$, and $c = 125$.

- Case V: Given the three sides, the law of cosines is used to find any angle

Solved Problem 10.7 Solve the triangle ABC, given $a = 25.2$, $b = 37.8$, and $c = 43.4$. See Figure 10-10.

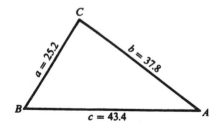

Figure 10-10

Solution.

For A: $\cos A = \dfrac{b^2 + c^2 - a^2}{2bc} = \dfrac{(37.8)^2 + (43.4)^2 - (25.2)^2}{2(37.8)(43.4)} = 0.8160$

and $A = 35°20'$

For B: $\cos B = \dfrac{c^2 + a^2 - b^2}{2ca} = \dfrac{(43.4)^2 + (25.2)^2 - (37.8)^2}{2(43.4)(25.2)} = 0.4982$

and $B = 60°10'$

For C: $\cos C = \dfrac{a^2 + b^2 - c^2}{2ab} = \dfrac{(25.2)^2 + (37.8)^2 - (43.4)^2}{2(25.2)(37.8)} = 0.0947$

and $C = 84°30'$

Check: $A + B + C = 180°$.

Chapter 11
AREA OF A
TRIANGLE

IN THIS CHAPTER:

✔ *Area of a Triangle*
✔ *Area Formulas*

Area of a Triangle

The area K of any triangle equals one-half the product of its base and altitude. In general, if enough information about a triangle is known so that it can be solved, then its area can be found.

Area Formulas

Cases I and II. Given two angles and a side of triangle *ABC*

The third angle is found using the fact that $A + B + C = 180°$. The area of the triangle equals a side squared times the product of the sines of the angles including the side divided by twice the sine of the angle opposite the side; i.e.,

$$K = \frac{a^2 \sin B \sin C}{2 \sin A} = \frac{b^2 \sin A \sin C}{2 \sin B} = \frac{c^2 \sin A \sin B}{2 \sin C}$$

Case III. Given two sides and the angle opposite one of them in triangle *ABC*

A second angle is found by using the law of sines and the appropriate formula from Case I. Since there are sometimes two solutions for the second angle, there will be times when the area of two triangles must be found.

Case IV. Given two sides and the included angle of triangle *ABC*

The area of the triangle is equal to one-half the product of the two sides times the sine of the included angle; i.e.,

$$K = (\tfrac{1}{2})ab \sin C = (\tfrac{1}{2})ac \sin B = (\tfrac{1}{2})bc \sin A$$

Case V. Given the three sides of triangle *ABC*

The area of a triangle is equal to the square root of the product of the semiperimeter and the semiperimeter minus one side times the semiperimeter minus a second side times the semiperimeter minus a third side; i.e.,

$$K = \sqrt{s(s-a)(s-b)(s-c)} \quad \text{where } s = \tfrac{1}{2}(a + b + c)$$

Solved Problem 11.1 Derive the formula $K = (\tfrac{1}{2})bc \sin A$. See Figure 11-1.

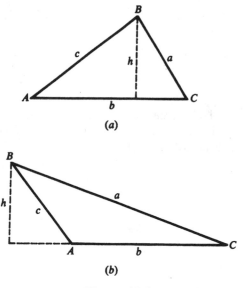

(a)

(b)

Figure 11-1

Solution. Denoting the altitude drawn to side b of the triangle ABC by h, we have from either figure $h = c \sin A$. Thus,

$$K = (\tfrac{1}{2})bh = (\tfrac{1}{2})bc \sin A$$

Solved Problem 11.2 Derive the formula

$$K = \frac{c^2 \sin A \sin B}{2 \sin C}$$

Solution. From Solved Problem 11.1, $K = (\tfrac{1}{2})bc \sin A$; and by the law of sines,

$$b = \frac{c \sin B}{\sin C}$$

Then,

$$K = (\tfrac{1}{2})bc \sin A = (\tfrac{1}{2}) \frac{c \sin B}{\sin C} c \sin A = \frac{c^2 \sin A \sin B}{2 \sin C} .$$

Solved Problem 11.3 A painter needs to find the area of the gable end of a house. What is the area of the gable if it is a triangle with two sides of 42.0 ft that meet at a 105° angle?

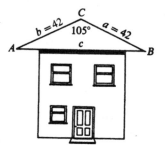

Figure 11-2

Solution. In Figure 11-2, $a = 42.0$ ft, $b = 42.0$ ft, and $C = 105°$.

$K = (\tfrac{1}{2})ab \sin C$

$\qquad = (\tfrac{1}{2})(42)(42) \sin 105°$

$\qquad = 852$ ft^2

Solved Problem 11.4 Three circles with radii 3.0, 5.0, and 9.0 cm are externally tangent. What is the area of the triangle formed by joining their centers?

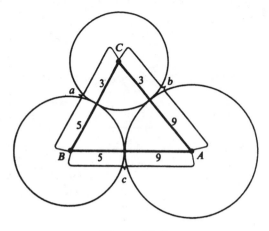

Figure 11-3

Solution. In Figure 11-3, $a = 8$ cm, $b = 12$ cm, and $c = 14$ cm.

$$s = \tfrac{1}{2}(a + b + c) = 17 \text{ cm}$$

$$K = \sqrt{s(s-a)(s-b)(s-c)}$$

$$= \sqrt{17(17-8)(17-12)(17-14)}$$

$$= \sqrt{2295}$$

$$= 48 \text{ cm}^2$$

Chapter 12
INVERSES OF TRIGONOMETRIC FUNCTIONS

IN THIS CHAPTER:

✔ *Inverse Trigonometric Relations*
✔ *Graphs of the Inverse Trigonometric Relations*
✔ *Inverse Trigonometric Functions*
✔ *Principal-Value Range*
✔ *General Values of Inverse Trigonometric Relations*

Inverse Trigonometric Relations

The equation

$$x = \sin y$$

defines a unique value of x for each given angle y. But when x is given, the equation may have no solution or many solutions. For example, if $x = 2$, there is no solution, since the sine of an angle never exceeds 1; if $x = \frac{1}{4}$, there are many solutions $y = 30°, 150°, 390°, 510°, -210°, -330°, \ldots$

$$y = \arcsin x$$

In spite of the use of the word *arc*, the above relation is to be interpreted as stating that "*y* is an angle whose sine is *x*." Similarly, we shall write $y = \arccos x$ if $x = \cos y$, $y = \arctan x$ if $x = \tan y$, etc.

The notation $y = \sin^{-1} x$, $y = \cos^{-1} x$, etc. (to be read "inverse sine of *x*, inverse cosine of *x*," etc.) are also used but $\sin^{-1} x$ may be confused with $1/\sin x = (\sin x)^{-1}$, so care in writing negative exponents for trigonometric functions is needed.

Graphs of the Inverse Trigonometric Relations

The graph of $y = \arcsin x$ is the graph of $x = \sin y$ and differs from the graph of $y = \sin x$ in that the roles of *x* and *y* are interchanged. Thus, the graph of $y = \arcsin x$ is a sine curve drawn on the *y* axis instead of the *x* axis.

Similarly, the graphs of the remaining inverse trigonometric relations are those of the corresponding trigonometric functions except that the roles of the *x* and *y* are interchanged.

Inverse Trigonometric Functions

It is at times necessary to consider the inverse trigonometric relations as functions (i.e., one value of *y* corresponding to each admissible value of *x*). To do this, we agree to select one out of the many angles corresponding to the given value of *x*. For example, when $x = \frac{1}{2}$, we shall agree to select the value $y = \pi/6$, and when $x = -\frac{1}{2}$, we shall agree to select the value $y = -\pi/6$. This selected value is called the *principal value* of arc-sin *x*. When only the principal value is called for, we shall write arcsin *x*, arccos *x*, etc. Alternative notation for the principal value of the inverses of the trigonometric functions is $\sin^{-1} x$, $\cos^{-1} x$, $\tan^{-1} x$, etc. The portions of the graphs on which the principal values of each of the inverse trigonometric relations lie are shown in Figure 12-1(*a*) to (*f*) by a heavier line.

When x is positive or zero and the inverse function exists, the principal value is defined as that value of y which lies between 0 and $(\frac{1}{2})\pi$ inclusive.

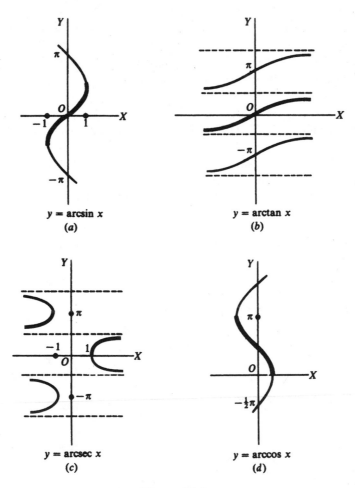

Figure 12-1

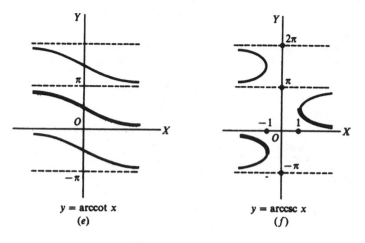

y = arccot x
(e)

y = arccsc x
(f)

Figure 12-1, cont.

Principal-Value Range

Authors vary in defining the principal values of the inverse functions when x is negative. The definitions given are the most convenient for the calculus. In many calculus textbooks, the inverse of a trigonometric function is defined as the principal-valued inverse and no capital letter is used in their notation. Since only the inverse function is considered, this generally causes no problem in the calculus class.

General Values of Inverse Trigonometric Relations

Let y be an inverse trigonometric relation of x. Since the value of a trigonometric relation of y is known, two positions are determined in general for the terminal side of the angle y. Let y_1 and y_2, respectively, be angles determined by the two positions of the terminal side. Then the totality of values of y consists of the angles y_1 and y_2, together with all angles coterminal with them, that is,

$$y_1 + 2n\pi \quad \text{and} \quad y_2 + 2n\pi$$

where n is any positive or negative integer or zero.

One of the values y_1 and y_2 may always be taken as the principal value of the inverse trigonometric function.

Solved Problem 12.1 Evaluate each of the following: (a) cos (arcsin 3/5), (b) sin[arccos (−2/3)], and (c) tan[arcsin (−3/4)].

Solution.

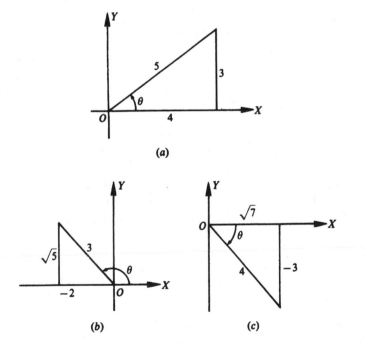

Figure 12-2

(*a*) Let $\theta = $ arcsin 3/5; then sin $\theta = $ 3/5, θ being a first-quadrant angle. From Figure 12-1(*a*),

$$\cos\,(\text{arcsin } 3/5) = \cos\,\theta\, = 4/5$$

(*b*) Let $\theta = $ arccos (−2/3); then cos $\theta = $ −2/3, θ being a second-quadrant angle. From Figure 12-1(*b*),

$$\sin\,[\text{arccos } (-2/3)] = \sin\,\theta\, = \sqrt{5}\,/\,3$$

(*c*) Let $\theta = $ arcsin (−3/4); then sin $\theta = $ −3/4, θ being a fourth-quadrant angle. From Figure 12-1(*c*),

$$\tan\,[\text{arcsin } (-3/4)] = \tan\,\theta\, = -3\,/\,\sqrt{7}\, = -3\sqrt{7}\,/\,7$$

Solved Problem 12.2 Evaluate sin (arcsin 12/13 + arcsin 4/5).

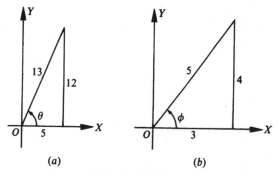

(*a*) (*b*)

Figure 12-3

Solution. Let

$$\theta = \arcsin 12/13$$

$$\phi = \arcsin 4/5$$

Then $\sin \theta = 12/13$ and $\sin \phi = 4/5$, θ and ϕ being first-quadrant angles. From Figure 12-2(a) and (b),

$$\sin (\arcsin 12/13 + \arcsin 4/5) = \sin (\theta + \phi)$$

$$= \sin \theta \cos \phi + \cos \theta \sin \phi$$

$$= \frac{12}{13} \cdot \frac{3}{5} + \frac{5}{13} \cdot \frac{4}{5} = \frac{56}{65}$$

Chapter 13
TRIGONOMETRIC EQUATIONS

IN THIS CHAPTER:

✔ *Trigonometric Equations*
✔ *Solving Trigonometric Equations*

Trigonometric Equations

Trigonometric equations, i.e., equations involving trigonometric functions of unknown angles, are called:

(*a*) Identical equation, or *identities*, if they are satisfied by all values of the unknown angles for which the functions are defined.

(*b*) Conditional equations, or equations, if they are satisfied only by particular values of the unknown angles.

Hereafter, in this chapter, we shall use the term *equation* instead of *conditional equation*.
A solution of a trigonometric equation, like $\sin x = 0$, is a value of the angle x which satisfies the equation. Two solutions of $\sin x = 0$ are $x = 0$ and $x = \pi$.

If a given equation has one solution, it has in general an unlimited number of solutions. Thus, the complete solution of $\sin x = 0$ is given by

$$x = 0 + 2n\pi \qquad x = \pi + 2n\pi$$

where n is any positive or negative integer or zero.

Solving Trigonometric Equations

There is no general method for solving trigonometric equations. Several standard procedures are illustrated in the following examples and other procedures are introduced in the Solved Problems below. All solutions will be for the interval $0 \le x < 2\pi$.

(A) The equation may be factorable.

Solved Problem 13.1 Solve $\sin x - (2 \sin x \cos x) = 0$.

Solution. Factoring, $\sin x - 2 \sin x \cos x = \sin x (1 - 2 \cos x) = 0$ and setting each factor equal to zero, we have

$$\sin x = 0 \qquad \text{and} \qquad x = 0, \pi$$

or $1 - 2 \cos x = 0$ and $\cos x = \frac{1}{2}$ and $x = \pi/3, 5\pi/3$

Check: For $x = 0$, $\sin x - 2 \sin x \cos x = 0 - 2(0)(1) = 0$

For $x = \pi/3$, $\sin x - 2 \sin x \cos x = \frac{1}{2}\sqrt{3} - 2(\frac{1}{2}\sqrt{3})(\frac{1}{2}) = 0$

For $x = \pi$, $\sin x - 2 \sin x \cos x = 0 - 2(0)(-1) = 0$

For $x = 5\pi/3$, $\sin x - 2 \sin x \cos x = -\frac{1}{2}\sqrt{3} - 2(-\frac{1}{2}\sqrt{3})(\frac{1}{2}) = 0$

Thus, the required solutions ($0 \leq x < 2\pi$) are $x = 0$, $\pi/3$, π, and $5\pi/3$.

(B) The various functions occurring in the equation may be expressed in terms of a single function.

Solved Problem 13.2 Solve $\sec x + \tan x = 0$.

Solution. Multiplying the equation

$$\sec x + \tan x = \frac{1}{\cos x} + \frac{\sin x}{\cos x} = 0$$

by $\cos x$, we have $1 + \sin x = 0$ or $\sin x = -1$; then $x = 3\pi/2$. However, neither $\sec x$ nor $\tan x$ is defined when $x = 3\pi/2$ and the equation has no solution.

(C) Both members of the equation are squared.

Solved Problem 13.3 Solve $\sin x + \cos x = 1$.

Solution. If the procedure of (B) were used, we would replace $\sin x$ by $\pm\sqrt{1 - \cos^2 x}$ or $\cos x$ by $\pm\sqrt{1 - \sin^2 x}$ and thereby introduce radicals. To avoid this, we write the equation in the form $\sin x = 1 - \cos x$ and square both members. We have

$$\sin^2 x = 1 - 2\cos x + \cos^2 x$$

$$1 - \cos^2 x = 1 - 2\cos x + \cos^2 x$$

$$2\cos^2 x - 2\cos x = 2\cos x\,(\cos x - 1) = 0$$

From $\cos x = 0$, $x = \pi/2$, $3\pi/2$; from $\cos x = 1$; $x = 0$.

Check: For $x = 0$, $\sin x + \cos x =$ $0 + 1 = 1$

$$\text{For } x = \pi/2, \qquad \sin x + \cos x = \qquad 1 + 0 = 1$$

$$\text{For } x = 3\pi/2, \qquad \sin x + \cos x = \qquad -1 + 0 \neq 1$$

Thus, the required solutions are $x = 0$ and $\pi/2$.

The value $x = 3\pi/2$, called an *extraneous solution*, was introduced by squaring the two members. Note that the equation

$$\sin^2 x = 1 - 2 \cos x + \cos^2 x$$

is also obtained when both members of $\sin x = \cos x - 1$ are squared and that $x = 3\pi/2$ satisfies this latter relation.

(D) Solutions are approximate values.

Solved Problem 13.4 Solve $4 \sin x = 3$.

Solution. From the equation $4 \sin x = 3$, we get

$$\sin x = {}^3\!/_4 = 0.75$$

The reference angle is 0.85 and the solutions for x are $x = 0.85$ and $x = \pi - 0.85 = 3.14 - 0.85 = 2.29$.

Check: For $x = 0.85$, $4 \sin 0.85 = 4(0.7513) = 3.0052 \approx 3$

For $x = 2.29$, $4 \sin 2.29 = 4[\sin (3.14 - 2.29)]$

$$= 4[\sin 0.85] = 4[0.7513] = 3.0052 \approx 3$$

If a calculator is used, $\sin 2.29$ is computed directly, so $4 \sin 2.29 = 4(0.7513) = 3.0092 \approx 3$. Thus, the solutions to the nearest hundredth radian are 0.85 and 2.29.

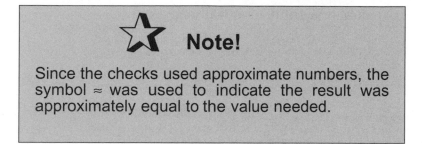

Note!

Since the checks used approximate numbers, the symbol $\approx$ was used to indicate the result was approximately equal to the value needed.

(E) Equation contains a multiple angle.

Solved Problem 13.5 Solve $\cos 2x - 3 \sin x + 1 = 0$.

Solution.

$$\cos 2x - 3 \sin x + 1 = 0$$

$$(1 - 2 \sin^2 x) - 3 \sin x + 1 = 0$$

$$- 2 \sin^2 x - 3 \sin x + 2 = 0$$
$$2 \sin^2 x + 3 \sin x - 2 = 0$$

$$(2 \sin x - 1)(\sin x + 2) = 0$$

and $\sin x = \frac{1}{2}$ and $\sin x = -2$.

From $\sin x = \frac{1}{2}$, $x = \pi/6$ and $5\pi/6$.

From $\sin x = -2$, there are no solutions since $-1 \leq \sin x \leq 1$ for all x.

The solutions for x are $\pi/6$ and $5\pi/6$.

(F) Equations containing half angles.

Solved Problem 13.6 Solve $4 \sin^2 (½)x = 1$.

Solution.

$4 \sin^2 (½)x = 1$

$\sin^2 (½)x = ¼$

$\sin (½)x = ±½$

Since we want $0 \leq x < 2\pi$, we want all solutions for $(½)x$ such that $0 \leq (½) x < \pi$.

From $\sin (½)x = ½$, $(½)x = \pi/6$ and $5\pi/6$, and $x = \pi/3$ and $5\pi/3$.

From $\sin (½)x = -½$, there are no solutions since $\sin (½)x \geq 0$ for all x such that $0 \leq (½)x < \pi$.

The required solutions are $x = \pi/3$ and $5\pi/3$.

Index